AGRICULTURE ISSUES AND POLICIES

BIOLOGICAL CONTROL

METHODS, APPLICATIONS AND CHALLENGES

Agriculture Issues and Policies

Additional books in this series can be found on Nova's website under the Series tab.

Additional e-books in this series can be found on Nova's website under the e-book tab.

AGRICULTURE ISSUES AND POLICIES

BIOLOGICAL CONTROL

METHODS, APPLICATIONS AND CHALLENGES

LEWIS DAVENPORT
EDITOR

NOTICE TO THE READER

Library of Congress Cataloging-in-Publication Data

ISBN: 978-1-53612-416-3

Published by Nova Science Publishers, Inc. † New York

CONTENTS

PREFACE

Biological control is a plant protection strategy widely-used in horticultural cropping systems to regulate insect and mite pest populations on greenhouse-grown ornamentals and vegetables. The use of natural enemies in controlling the SWD has been researched in Chapter One. Chapter Two aims to recover the history and present the current status of the biological control of fruit flies of the genus Anastrepha and to show perspectives of the use of such method. Chapter Three discusses the advantages and issues affiliated with using natural enemies in conjunction with pesticides and provides insights on the practicality of using both plant protection strategies simultaneously. Chapter Four describes screening methods for candidate biocontrol agents, application methods and challenges associated with biological control of fungal pathogens by antagonistic microorganisms.

Chapter 1 - *Drosophila suzukii* (Diptera: Drosophilidae), also known as Spotted Wing Drosophila (SWD) is a species originated in Japan and rapidly spread throughout Europe and North America since 2007. In 2013, it was found in southern Brazil, and nowadays there are records in other South American countries. This specie of Drosophilidae is already considered a pest to small fruits and stone fruits in Asia, Europe, and North America. *Drosophila suzukii* females are capable of introducing the ovipositor inside intact fruits, allowing the entrance of phytopathogenic organisms and rendering fruits unviable due the development of immature phases within the pulp. Furthermore, the species presents a high biotic potential, as it may reach elevated population levels in a short period, whenever climatic conditions are favorable. Currently, the SWD control is mostly accomplished by means of chemical control, with coverage applications or the use of toxic baits. However, because of restrictions to the use of synthetic insecticides in target crops, the risk of resistant lineages arising, and the negative effect over non-target organisms, such as pollinators and natural enemies, the development of alternative management techniques is essential. Amongst the alternatives, the biological control has been the most studied one, and it will be approached in this chapter. The use of natural enemies in controlling the SWD has been researched in several parts of the world. Up until now, several parasitoid species were found to be associated to *D. suzukii* demonstrating the possibility of application in future integrated management programs. Moreover, some studies have already been made as to explore the efficiency of other natural enemies, such as predators, bacteria, fungi, and nematodes, which have displayed a potential for the insect's population suppression. The biological control technique presents great potential for SWD management, and it can be associated to other methods, such as chemical, autocidal (Sterile Insect Technique – SIT, Release of Insects with Dominant Lethality – RIDL and Wolbachia-induced Incompatible Insect Technique - IIT), and cultural controls, which not only do not influence other agronomic treatments, but are also environmentally sustainable. In face of the current scenario, efforts expected to manage this species lean towards a more integrated control, wherein biological control may become an important tool in the near future, thus minimizing the negative effect caused by the reliance in chemical control.

Chapter 2 - Historically, control of fruit flies has been performed with insecticides applied for full coverage. The use of synthetic insecticides in orcharding has aroused greater care due to market requirements, human and animal health problems and environmental impacts. In this context, the biological control of fruit flies becomes essential, since it aims to control the populations of the pest in balance with the new environmental, health and economic requirements. Biological control of fruit flies can be done mainly with the use of parasitoids, however the use of pathogens (bacteria, fungi and nematodes) and predators can be promising from an Integrated Pest Management perspective. In view of the above, this chapter aims to recover the history and present the current status of the biological control of fruit flies of the genus *Anastrepha* and to show perspectives of the use of such method.

Chapter 3 - Biological control is a plant protection strategy widely-used in horticultural cropping systems to regulate insect and mite pest populations on greenhouse-grown ornamentals and vegetables. However, in certain instances, biological control may not provide sufficient regulation of insect and mite pest populations. Therefore, pesticides (insecticides and miticides) are oftentimes required to provide supplemental suppression of pest populations. The integration of natural enemies or biological control agents (parasitoids and predators) and pesticides is a strategy that is gaining interest among greenhouse producers due to the potential for reduced pesticide inputs, which may result in decreased development of pesticide resistance in insect and mite pest populations, and potential economic and environmental benefits. This chapter discusses the advantages and issues affiliated with using natural enemies in conjunction with pesticides and provides insights on the practicality of using both plant protection strategies simultaneously. Furthermore, the direct and indirect effects of entomopathogenic fungi, botanical pesticides, insecticidal soaps and horticultural oils, and fungicides on natural enemies are discussed. Finally, guidelines for enhancing the integration of biological control and pesticides are presented.

Chapter 4 - Worldwide plant disease control usually relies on the use of fungicides. However, pathogen resistance to fungicides, developing after frequent treatments, and host sensitivity to fungicides have been recognized as serious problems. Disease control with only a few or no chemicals available at all is a major challenge for growers in the 21st century. Considering pathogen resistance evolution and harmful impact on the environment and human health, special attention has been focused on biofungicides based on antagonistic microorganisms. Previous studies have indicated that various biological agents appear to be suitable, eco-friendly alternatives as they have demonstrated strong antifungal effects against pathogenic fungi on cultivated plants and mushrooms. Microbiological products are usually based on antagonistic microorganisms, such as bacteria (*Bacillus* spp., *Lactobacillus* spp. and *Pseudomonas* spp.), fungi (*Pythium* spp. and *Trichoderma* spp.) and actinomycetes (*Streptomyces* spp., etc.). The use of microbial inoculants registered for biological control applications is the result of their ability to antagonize the pathogen by multiple modes of action and also to effectively colonize the rhizosphere or phyloshere. Strong activities of antagonistic species are due to their production of antibiotics, volatile compounds, lytic enzymes, and plant growth-promotion effects by phytohormones. The efficacy of antagonistic organisms depends on many abiotic and biotic factors, climatic conditions and interactions with other soil-inhabiting organisms. The introduction of biofungicides has created new possibilities for crop protection with reduced application of chemicals.

In: Biological Control
Editor: Lewis Davenport
ISBN: 978-1-53612-416-3

Chapter 1

BIOLOGICAL CONTROL OF *DROSOPHILA SUZUKII* (DIPTERA, DROSOPHILIDAE): STATE OF THE ART AND PROSPECTS

Flávio Roberto Mello Garcia[1,*], Jutiane Wollmann[1], Alexandra Peter Krüger[1], Daniele Cristine Hoffmann Schlesener[1] and Cristiano Machado Teixeira[1]
[1]Federal University of Pelotas, Pelotas, Brazil

ABSTRACT

Drosophila suzukii (Diptera: Drosophilidae), also known as Spotted Wing Drosophila (SWD) is a species originated in Japan and rapidly

* Corresponding Author. Department of Ecology, Zoology and Genetics, Insect Ecology Lab, UFPEL, Institute of Biology, Pelotas, RS, Brazil. Email flaviormg@hotmail.com

spread throughout Europe and North America since 2007. In 2013, it was found in southern Brazil, and nowadays there are records in other South American countries. This specie of Drosophilidae is already considered a pest to small fruits and stone fruits in Asia, Europe, and North America. *Drosophila suzukii* females are capable of introducing the ovipositor inside intact fruits, allowing the entrance of phytopathogenic organisms and rendering fruits unviable due the development of immature phases within the pulp. Furthermore, the species presents a high biotic potential, as it may reach elevated population levels in a short period, whenever climatic conditions are favorable. Currently, the SWD control is mostly accomplished by means of chemical control, with coverage applications or the use of toxic baits. However, because of restrictions to the use of synthetic insecticides in target crops, the risk of resistant lineages arising, and the negative effect over non-target organisms, such as pollinators and natural enemies, the development of alternative management techniques is essential. Amongst the alternatives, the biological control has been the most studied one, and it will be approached in this chapter. The use of natural enemies in controlling the SWD has been researched in several parts of the world. Up until now, several parasitoid species were found to be associated to *D. suzukii* demonstrating the possibility of application in future integrated management programs. Moreover, some studies have already been made as to explore the efficiency of other natural enemies, such as predators, bacteria, fungi, and nematodes, which have displayed a potential for the insect's population suppression. The biological control technique presents great potential for SWD management, and it can be associated to other methods, such as chemical, autocidal (Sterile Insect Technique – SIT, Release of Insects with Dominant Lethality – RIDL and Wolbachia-induced Incompatible Insect Technique - IIT), and cultural controls, which not only do not influence other agronomic treatments, but are also environmentally sustainable. In face of the current scenario, efforts expected to manage this species lean towards a more integrated control, wherein biological control may become an important tool in the near future, thus minimizing the negative effect caused by the reliance in chemical control.

Keywords: invasive species, natural enemies, spotted wing drosophila

INTRODUCTION

The Drosophilidae (Diptera) family is formed by several species occupying various ecological niches distributed worldwidely. Originally,

Drosophila larvae most likely fed on decomposing and fermenting vegetation. A change regarding the dependency on fermentation-associated microorganisms, such as yeast and bacteria, has allowed Drosophila to expand their diet towards other resources, such as fruit, sap, and fungi, all of which support the growth of certain microorganisms (Powell 1997). This behavior causes this group of insects not to be elevated to a pest status, since they do not compete with humans for the same resource.

However, contrasting with the habit presented by most drosophilids, the *Drosophila suzukii* (Matsumura, 1931) (Diptera: Drosophilidae) has the capability to infest intact fruit, piercing them in order to deposit eggs and, since the species was found in 2008 in a Californian raspberry field (Hauser 2009; 2011), it has expanded rapidly throughout several territories and it has been causing great damage to a variety of economically significant hosts.

Commonly known as the spotted wing drosophila (SWD), this species originated in Asia and it was first detected outside its original environment in 1980 in Hawaii and, afterwards, in other Hawaiian islands, there been no records of any damages (Kaneshiro 1983; O'Grady et al. 2002). In 2008, it was recorded simultaneously in California, United States (Hauser 2009; 2011) and many European localities (Calabria et al. 2012). In South America, it has been recorded simultaneously in Brazil (Deprá et al. 2014; Schlesener et al. 2014; Geisler et al. 2015) and in Uruguay (González et al. 2015). Its rapid dispersion across western countries has demonstrated this species' capability for adapting to newly invaded areas, where massive losses have been recorded in agriculture (Bolda et al. 2010; Goodhue et al. 2011; Lee et al. 2011; De Ros et al. 2013). A recent study indicates environmentally adequate areas with particular predisposition to the occurrence of *D. suzukii* in Oceania and Africa, although there have been no record of the species occurring in those continents. Models indicate that environmental conditions in those areas are prone to the establishment of said species in case of future invasion (Dos Santos et al. 2017).

Some hypothesis ought to be considered regarding dispersion and colonization of SWD in invaded areas. Human activity – particularly the international trade of infested fruits – is most likely the means by which said species has spread through several continents (Westphal et al. 2008), as well as climate conditions similar to the original location (Wiman et al. 2014). Another factor relates to the "Enemy Release Hypothesis" (ERH) theory, which tries to explain the establishment and proliferation of invading species. According to this theory, exotic species introduced to a new territory will suffer less of an impact from natural enemies, which leads to an increase and distribution of the invading species. This happens because natural control agents have a stronger regulating effect over indigenous species in detriment to exotic ones, and this difference in regulation results in an increased population density of invading species (Keane & Crawley 2002; Roy et al. 2011).

The *Drosophila suzukii* species has a short biological cycle, with high levels of reproduction and generation overlapping (Emiljanowicz et al. 2014; Tochen et al. 2014), considered today to be one of the main pests for small and stone fruit (Asplen et al. 2015). Its ability to lay eggs in intact mature or maturing fruit is attributed to the shape of its ovipositor, narrow and serrated, with a row of robust, sclerotized teeth (Walsh et al. 2011; Anfora et al. 2012; Daff 2013). The insects' larvae develops in the fruit tissue causing its softening, and the orifice formed by oviposition allows microorganisms to enter, leading to decay, accelerating decomposition and making fruits improper for commercial purposes (Walsh et al. 2011).

Hosts that are potentially susceptible to SWD attack are, mostly, blackberries, blueberries, raspberries, and cherries (Bolda et al. 2010; Lee et al. 2011; 2015), but the fly may also infest kakis, figs, apples, pears, nectarines, loquats, grapes, and others (Lee et al. 2011; Walsh et al. 2011). Regarding other, economically unimportant hosts, in Europe and in the United States over 50 species have been recorded as such (amongst wild fruits and ornamental plants), providing alternate hosting in the absence of commercial crops (Lee et al. 2015; Poyet et al. 2015; Kenis et al. 2016). The wide range of hosts and the ability to damage ripening fruit allow the SWD to exploit an economical niche that is both broad and specialized (Cini et al. 2012).

Currently, chemical control is the main method employed in *D. suzukii*'s management. Nevertheless, since this insect is able to produce many generations during one crop season, and because insecticide control has limited residual power, frequent enforcements are required in order to reduce the pests' population levels (Bruck et al. 2011). This frequency in applications may increase both the risk that maximum insecticide residue limits are exceeded and the potential for resistance development. (Renkema et al. 2014).

Other methods – either combined or not with chemical control – have been recommended at the ripening stage. Maintaining orchards healthy can be accomplished via cultural control, by removing fallen and over-ripe fruits, and wild hosts surrounding the farming areas (Cini et al. 2012; Walsh et al. 2011). Physical control by covering plants with fine mesh nets can also reduce infestations in orchards, because of the pest exclusion principle (Leach et al. 2016). Another control method gaining relevance is the biological control as a regulating factor in pest organism's population.

Regarding biological insect control, which includes a complex of entomophagus organisms, such as parasitoids and predators, the entomopathogens – as bacteria, nematodes, and fungi – numerous researches have been developed all over the world seeking biotic mortality agents that are effective against the SWD. Natural enemies can proliferate in commercial crops as well as in natural habitats, playing a unique role in insect population regulation (Haye et al. 2016), to an extent that, in recently invaded areas, those agents can gradually adapt and establish new associations, and thus doing contribute to a future suppression of pest populations (Rossi-Stacconi et al. 2013). In this context, the biological control strategy can be economically viable, environmentally safe, and employed alongside other methods in *D. suzukii*'s integrated management control.

Therefore, given the economic importance of the crops targeted by *D. suzukii* and the elevated impact that this pest has had worldwide, new strategies addressing its control are a necessity and a challenge to be overcome in the short run. Consequently, this chapter is dedicated to emphasize, in the light of current knowledge, the efforts made in order suppress SWD by means of biological control, which may become an important tool in the near future, be it by means of conservation methods aimed at natural enemies existing in the invaded areas, by classical biological control, or augmentative control.

PARASITOIDS

Parasitoids are a large group of insect, formed particularly by specimens from the Hymenoptera and Diptera orders. These insects develop within or on the surface of another subject's body (usually another insect) and consume the hosts' tissue during their development. Parasitoids play an important role in the insect community, acting as a population-controlling agent – either naturally or through biological control programs (Chabert et al. 2012) – as well as acting as a useful bio-indicator of ecosystem quality and diversity (Heavner et al. 2014). Amongst the *D. suzukii* biological control agents, parasitoids are the most commonly studied and the ones presenting the upmost probability for success.

Every continent, innate or invaded by *D. suzukii* has recorded the occurrence of parasitoids in larvae or pupae associated to the species. Nevertheless, until now only generalist parasitoids have been found to parasitize SWD. In order to use parasitoids as biological control agents, it is preferable that they are specialized on the pest that one intends to fight, since generalist parasitoids can affect other species that one does not intend to control (Nomano et al. 2015).

Several species, from different families and parasitoid larvae genus have been recorded for *D. suzukii* in different parts of the world. The *Asobara* (Hymenoptera: Braconidae) genus presents the most species associated to the SWD. *Asobara japonica* Belokobylkij, 1998 and *A. leveri* (Nixon, 1939) have been verified to associate to *D. suzukii* in Asia (Mitsui et al. 2007; Mitsui & Kimura 2010; Kasuya et al. 2013; Daane et al. 2016; Guerrieri et al. 2016), *A. tabida* (Nees, 1834) in Asia (Mitsui et al. 2007) and Europe (Chabert et al. 2012), as well as five new species, *A. elongata*, *A. mesocauda*, *A. unicolorata* and *A. triangulata* van Achterberg & Guerrieti, 2016 (Hymenoptera: Braconidae) found in China, and *A. brevicauda* van Achterberg & Guerrieti, 2016 (Hymenoptera: Braconidae) found in South Korea, that can potentially parasitoid *D. suzukii* (Guerrieri et al. 2016). One species from that same genus (*Asobara* sp. TK1) was found in Japan to have great potential as a biological control agent for *D. suzukii*, since it has emerged only pupae from that species (Nomano et al. 2015). Two species from the *Ganaspis* (Hymenoptera: Figitidae) genus, *G. brasiliensis* (Ihering, 1905) (Daane et al. 2016) and *G. xanthopoda* (Ashmead, 1896) were verified in association with *D. suzukii* in the Asian continent alone.

The *Leptopilina* (Hymenoptera: Figitidae) genus is also formed by larvae parasite and it was widely found in association with *D. suzukii.* The *L. japonica* Novkovic & Kimura, 2011 species was confirmed in the Asian continent alone. However, other species from this genus were found in other continents invaded by this pest, such as *L. heterotoma* (Thomson, 1862) in Europe and North America (Chabert et al. 2012; Miller et al. 2015; Rossi-Stacconi et al. 2015) and *L. boulardi* (Barbotin, Carton & Kelner-Pillault, 1979), found in Europe, North America, and South America (Chabert et al. 2012; Cuch-Arguimbau et al. 2013; Cancino et al. 2015; Wollmann et al. 2016).

Despite the expressive number of larvae parasitoid species found in association with *D. suzukii*, few subjects manage to complete its biological cycle with this species as host. This fact is, mainly, due the complex immune system and to the great number of hemocytes present in the species. Hemocytes are cells that constitute blood, and in insects they form the last barrier in their immune system, acting in the coagulation and scarring, in phagocytosis, encapsulation, and nodulation, as well as antimicrobial factors like the induced synthesis of peptides and antimicrobial proteins (Lavine & Strand 2002). *Drosophila suzukii* presents the highest number of circulating hemocytes ever recorded in *Drosophila* larvae within the *melanogaster* (Poyet et al. 2013).

Many cell types are involved in the defense process and immune system of drosophilids. Plasmocytes are the most abundant type of hemocytes circulating in the hemolymph. They are responsible for the phagocytosis of small foreign particles, recognizing wasp eggs, inducing the differentiation of lamellocytes, and participating in the final processes of encapsulation. Crystal cells are the rarest type of hemocytes in the hemolymph, and it contains elements from the phenoloxidase cascade (substrate and enzymes) that are involved in the process of capsule melanization. Lamellocytes are themselves large cells produced as a response to parasitic infections, responsible for the encapsulation of those larger foreign bodies that cannot be phagocytized, like wasp eggs (Krzemien et al. 2010; Poyet et al. 2013). After encapsulation, the wasp egg is in direct contact with internal capsule cell that release reactive oxygen, forming an impermeable layer of melanin and leading to the death of the egg. However, the presence of a large amount of hemocytes alone does no guaranty a good immune response, as it is necessary to differentiate lamellocytes for the process of encapsulation to occur (Kacsoh & Schlenke 2012).

The great success of pupal parasitoids (pupa-pupa) regarding *D. suzukii* is probably attributed to an increased vulnerability the larvae present in this phase. A diminished immune response from the pupae can also explain the more generalist behavior from parasitoids that use the insects' biological phase as substrate for oviposition (Kacsoh & Schlenke 2012). Basically, two pupa-pupa parasitoid genus have been recorded worldwidely associated to SWD, *Trichopria* (Hymenoptera: Diapriidae) and *Pachycrepoideus* (Hymenoptera: Pteromalidae).

Species from the *Trichopria* genus are pupal parasitoids that oviposit within the host's hemocele. *Trichopria drosophilae* (Perkins, 1910) or *T.* cf *drosophilae* (cf = requires confirmation) has been verified associated to the SWD in Asia, Europe, and North America (Gabarra et al. 2015; Cancino et al. 2015; Miller et al. 2015; Rossi-Stacconi et al. 2015; Daane et al. 2016; Wang et al. 2016). In South America the occurrence of *T. anastrephae* Lima, 1940 parasiting *D. suzukii* has been recorded (Wollmann et al. 2016).

The *Pachycrepoideus vindemmiae* (Rondani, 1875) is an ectoparasitic pupal parasitoid (Carrillo et al. 2015). Its occurrence associated to *D. suzukii* has been verified in Asia, Europe, and North America (Brown et al. 2011; Chabert et al. 2012; Cuch-Arguimbau et al. 2013; Rossi-Stacconi et al. 2013; 2015; Cancino et al. 2015; Miller et al. 2015; Daane et al. 2016).

This generalist parasitoid is able to attack species from different insect orders, such as Diptera, Hemiptera, Hymenoptera, and Lepidoptera. Furthermore, it is considered a facultative hyperparasitoid, in other words, it is able to parasite other primary parasitoids, such as *Diachasmimorpha fullawayi* (Silvestri, 1913), *D. tryoni* (Cameron, 1911), *Psyttalia humilis* (Silvestri, 1913) (Hymenoptera: Braconidae), *Coptera silvestrii* (Kieffer, 1913) (Hymenoptera: Diapriidae), *Tetrastichus giffardianus* Silvestri, 1915 (Hymenoptera: Eulophidae), in addition to other parasitoids that integrate biological control programs for pests in the Tephritidae family, like the braconids *Fopius arisanus* (Sonan, 1932), *Diachasmimorpha longicaudata* (Ashmead, 1905), *D. kraussii* (Fullaway, 1951) and *Psyttalia concolor* (Szépligeti, 1910) (Wang & Messing 2004). Due to its low specificity and hyperparasitism, this parasitoid is not suitable to integrate a *D. suzukii* management program.

The lack of efficient natural autochthonous enemies in the invade areas may be contributing to a rapid dispersion and establishment of *D. suzukii* (Gabarra et al. 2015). Many works in probable origin areas of the species have been developed, in order to find parasitoids that are passible of quarantine evaluation and eventual field release, according to the precepts of classical biological control (Guerrieri et al. 2016). It is believed that parasitoids from the pest's place of origin can be more effective, because of the host–parasite coevolution (Daane et al. 2016).

Another contributing factor when it comes to the dispersion of *D. suzukii* is the lack of niche competition for reproducing and feeding, since this is the only *Drosophila* species to use ripening intact fruits as oviposition substrate. This increase in energy can be relocated to, amongst other things, hemocytes production and a sub consequent increase in parasitoid resistance (Poyet et al. 2013).

PREDATORS

Although efforts in the pursuit of *D. suzukii* biological control agents mainly focusing on parasitoids, some studies describe the impact predators have in regulating the population of other fruit fly species (Allen & Hagley 1990; Hodgson et al. 1998; Urbaneja et al. 2006; Monzo et al. 2011; Picchi et al. 2017). In fact, despite the protection from predators that the eggs and the first instars of SWD have when inside the fruit, the larval development, particularly in elevated densities, causes rapid fruit decay, and favors the exposure of larvae and pupae to predators (Renkema et al. 2015).

Until now, some predator insect species have been associated to plants infested by *D. suzukii*: *Orius insidiosus* (Say, 1832) in blueberries, *O. laevigatus* (Fieber, 1860) (Hemiptera: Anthocoridae) in strawberries and raspberries, *Cardiastethus nazarenus* Reuter 1884 and *C. fasciventris* (Garbiglietti, 1869) (Hemiptera: Anthocoridae) in strawberries and *Arbutus unedo* Linnaeus fruits, *Dicyphus tamaninii* Wagner, 1951 (Hemiptera: Miridae) in *Solanum luteum* Miller fruits, and *Labidura riparia* (Pallas, 1773) (Dermaptera: Labiduridae) in raspberries (Arnó et al. 2012; Gabarra et al. 2015; Woltz et al. 2015). Other than those, some species have been tested only in lab conditions: *O. majusculus* (Reuter, 1879), *Anthocoris nemoralis* (Fabricius 1794) (Hemiptera: Anthocoridae), *Dalotia coriaria* (Kraatz, 1856) (Coleoptera: Staphylinidae), *Forficula auricularia* (Linnaeus, 1758) (Dermaptera: Forficulidae), *Chrysoperla carnea* (Stephens, 1836) (Neuroptera: Chrysopidae) and *Hypoaspis miles* Berlese, 1892 (Acari: Laelapidae) (Cuthbertson et al. 2014a; Renkema et al. 2015; Woltz et al. 2015; Englert & Herz 2016).

Insects from the *Orius* genus are polyphagous predators and they are sold in some countries for the control of thrips and aphids (Funderburk et al. 2000, Rutledge & O'Neil 2005). In lab tests, *O. laevigatus* females did not prey on *D. suzukii* larvae, although the predator females were indeed able to diminish significantly the number of eggs by the pest (Gabarra et al. 2015). In another study *O. laevigatus* preyed upon adult SWD subjects, showing a preference for the males, whilst *O. majusculus* preferred to prey upon adult females (Cuthbertson et al. 2014a). *Orius majusculus* has not yet been observed preying upon de *D. suzukii* eggs and larvae (Englert & Herz 2016). Also in lab tests, *O. insidiosus* has been able to reduce the immature population of SWD between 12 and 50%, however when exposed to semi-field conditions, the pest population reduction has not been sufficient as to register a significant difference, showing a decline in preying capacity with an increase in system complexity (Woltz et al. 2015). The *A. nemoralis* was also observed feeding on SWD adults in laboratory, leading to a 45% mortality after five days exposure (Cuthbertson et al. 2014a). Insects from the *Cardiastethus* genus have already been recorded as predators to *Ceratitis capitata* (Wiedman, 1824) (Diptera: Tephritidae) eggs in citrus, and its presence in strawberry orchards infected by *D. suzukii* suggests that they are able to act on the species's biological control, although lab confirmation is still pending. Another predator from the hemipterus order associated to SWD infested orchards is the *D. tamaninii*, a polyphagous arthropod predator that was incorporated to biological control programs for pests of some vegetable species, despite lab tests still being necessary in order to verify the specie`s capacity for preying upon *D. suzukii* (Arnó et al. 2012).

Dalotia coriaria is a small predator that inhabits the soil and is used in the control of some dipterous and thrips species, mostly in sheltered sites (Jandricic et al. 2006; Birken & Cloyd 2007; Pochubay et al. 2015). Some studies evaluating the predation of *D. suzukii* larvae and pupae by the aforementioned predator do not display considerable effects on the reduction of pest population, suggesting that the predator is not able to access the larvae inside intact fruits (Woltz et al. 2015; Cuthbertson et al. 2014a). On the other hand, whenever larval density is high, rupturing of the fruit tissues allows for a greater exposure of immature forms to the predators, resulting in a decrease of up to 50% in fly emergence, with consumption rate of about 0.5 larvae per beetle per day (Renkema et al. 2015).

In laboratory studies, *L. riparia* nymphs and adults captured by traps located in *D. suzukii* infested orchards were observed consuming a large quantity of SWD larvae and pupae, indicating a capability for collaboration regarding insect population reduction (Gabarra et al. 2015). In addition, *L. riparia*is able to climb plants up to 35 cm in order to reach its prey, and that can be favorable when trying to get to the infested fruits still attached to the plant (Ammar & Farrag 1974; Gabarra et al. 2015). Yet another dermapterous species, *F. auricularia* has been observed preying upon *D. suzukii* larvae and pupae in lab conditions (Englert & Herz 2016).

Another experiment performed in laboratory have also described that the *C. carnea*) predator was able to prey upon *D. suzukii* in all developing phases, however the efficiency percentage was not disclosed (Englert & Herz 2016). The predator mite *H. miles* has been tested in *D. suzukii* larvae and pupae, but no sign of consumption was observed (Cuthbertson et al. 2014a).

ENTOMOPATHOGENS

The ecological relationships existing between insects and microorganisms may reveal themselves as highly informative in the employment of management strategies for pests granted the pest status. Some of those strategies consist in using entomopathogens such as bacteria, nematodes, and fungi.

Bacteria

Bacteria from the *Bacillus thuringiensis* Berliner 1915 (Bacillales: Bacillaceae) complex, also known as *Bt*, have been used as effective biological control agents against mosquito larvae and other insect groups. Their polypeptides grant them a larvicidal activity, separated in large (134, 128, 72 and 27 kDa) and small unities (78 and 29 kDa), respectively coded by cry4Aa, cry4Ba, cry11Aa, cyt1Aa, cry10Aa and cyt2Ba. They are highly specific concerning their action, and they act rupturing the membranes from the insects' digestive system. The lack of significant resistance levels in mosquito populations treated with these bio insecticides for decades suggests that they may become a biocontrol agent in a few years' time (Ben-Dov 2014).

Considering the promising field of application for Bt, studies have been performed with bacterial lineages against *D. suzukii* larvae and adults.

By testing 22 strains of *Bacillus thuringiensis* var. *thuringiensis*, *kurstaki*, *thompsoni*, *bolivia* and *pakistani*, researchers have observed results as high as 75 to 100% mortality in the first instar. However, with no effect on pupae mortality, that is, no interference in adult emergence. The larval state most susceptible to *B. thuringiensis* var. *kurstaki* was the first one, nevertheless, due to the insect's behavior (its feeding on fruit endocarp), larvae would be protected from contact with the bacteria. Thus, adults would be more vulnerable to *B. thuringiensis*. Authors have tested sucralose suspensions with *B. thuringiensis* var. *thuringiensis* strains, prepared for ingestion by *D. suzukii*, which lead to a low mortality in seven days. According to the authors, this would not be a good candidate for use in pest management, for it produces β-exotoxin, toxic to vertebrates (Cossentine et al. 2016a).

Nematodes

Nematodes pose as natural enemies for several insect species that have at least one phase of their life cycle associated to soil. Flies in general, and drosophilids that infest fallen fruits in particular, spend at least one phase as pupae on the soil or close to it, where some degree of controlling these insects naturally may be possible. However, since *D. suzukii* infests mostly intact fruits straight from the plant, the action of entomopathogenic nematodes is restricted, occurring occasionally when infested fruit falls to the ground.

Accordingly, the isolated nematodes *Steinernema carpocapsae* (Weiser, 1955), *S. feltiae* (Filipjev, 1934) and *S. kraussei* (Steiner, 1923) (Rhabditida: Steinernematidae) were evaluated, resulting in a reduction of SWD population development. Notwithstanding, it was determined that the tested entomopathogens seem unable to provide any control over *D. suzukii*, especially when used separately (Cuthbertson et al. 2014b).

Another work performed by Woltz et al. (2015) tested *Heterorhabditis bacteriophora* Poinar, 1976 (Rhabditida: Heterorhabditidae), *S. feltiae* and

S. carpocapsae. Like in the previous work, the tested nematodes presented low rates of infection and they were unable to affect *D. suzukii* survival and, although some limited control capability has been shown in lab conditions, these enemies are naturally present in farming soil, where they can contribute to the suppression of *D. suzukii*.

Fungi

Fungi represent some of the most commonly used natural enemies in suppressing insect population growth. Various means of employing these agents can be used. Some lab studies aim to test the form of action by these organisms in different application types. Therefore, short term and life table experiments were conducted continuously, as to determinate the impact of three fungi species over the growth and development of *D. suzukii*. Strains tested to *Geotrichum candidum* Link (1809) (Saccharomycetales: Endomycetaceae), *Talaromyces minioluteus* (Dierckx, 1901) (Eurotiales: Trichocomaceae) and *Actinomucor elegans* (Eidam) C. R. Benj. & Hesselt. 1957 (Mucorales: Mucoraceae) species. Results show that *T. minioluteus* had a negative effect on SWD survival. In addition, it was suggested that *G. candidum* and *A. elegans* have a mutualistic association between them and the *D. suzukii*, making them attractant specific baits, which could be mixed with traditional baits to a certain proportion and employed in mass capture (Gao et al. 2017).

Lab tests with commercial products based on fungi, regarding direct toxicity effects by products from *Beauveria bassiana* (Bals-Criv.) Vuill. (1912) (Hypocreales: Clavicipitaceae) have shown some efficacy in *D. suzukii* mortality (Gargani et al. 2013).

The susceptibility of adult SWD subjects was evaluated in lab tests using the Pf21, Pf17, Pf15 lineages of *Isaria fumosorosea* (Wize, 1904) (Hypocrales: Cordycipitaceae) and the Ma59 lineage of *Metarhizium anisopliae* (Metschn.) Sorokin 1883 (Hypocreales: Clavicipitaceae). Results show that the Pf21 strain of *I. fumosorosea* caused the highest mortality percentage (85%) from mycoses, followed by the Pf17 (60%),

Pf15 (57%) and *M. anisopliae* (Ma59) with 12%, clearly showing the real possibility for using these fungi in pest control management (Naranjo-Lázaro et al. 2014).

In another study, authors evaluated the effect of *M. anisopliae* Sor. (Ma198) and the teleomorphic phase of *B. bassiana,* [*Cordyceps bassiana* Z.Z. Li, C.R. Li, B. Huang & M.Z. Fan 2001 (Cb249)] in controlling *D. suzukii* adults in blackberry crops. With a conidia dose of $2x10^{13}$ /ha. The Cb and Ma strains controlled 60.3 and 37% of adult density and they also reduced the percentage of infested fruit in 71.1 and 57% respectively. Thus concluding that the Cb249 does have a potential as control agent for *D. suzukii* on the field (Peralta-Manzo et al. 2014).

In the work conducted by Woltz et al. (2015) *M. anisopliae*, *B. bassiana* and *I. fumosorosea* fungi were tested against *D. suzukii* adults and larvae. Only the *M. anisopliae* significantly diminished the pest's survival, but due to the low residual activity, no effect over the fly's fecundity was verified. Despite the limited capacity to suppress the SWD, the tested fungi are naturally present as a part of the community endemic to agricultural environments, where they are able to contribute in the suppression of *D. suzukii* (Woltz et al. 2015).

In another study, the entomopathogenic agents *Lecanicillium muscarium* (Petch) Zare & W. Gams 2001 (Hypocreales: Cordycipitaceae) and *B. bassiana* were tested, however results were not very expressive and the *B. bassiana* caused mortality to 44% of the subjects (Cuthbertson et al. 2014b).

Exposure of *D. suzukii* adults over surfaces with de *Metarhizium brunneum* Petch 1935 (Hypocreales: Clavicipitaceae), *B. bassiana*, *I. fumosorosea* and *Lecanicillium lecanii* (Zimm.) Zare & W. Gams 2001 (Hypocreales: Cordycipitaceae) conidia coating caused infection on the flies and mortality depending on the dosage. Approximately 50% of the flies died after exposure to $1x10^8$ conidia of each, isolated for 7, 10, 12, and 13 days respectively. It was possible to ascertain that the oviposition data was consistently lower, which indicates that the fungi had a negative effect over the SWD fecundity (Cossentine et al. 2016b).

Fungi from a *M. brunneum*, H.J.S. 1154, and *B. bassiana* strains were tested against pupae deposited over conidia inoculated soil, by direct application and immersion of the pupae in conidia suspension. This technique was not successful, since authors suggest that the pupal stage probably is too brief to allow entomopathogens to cause significant reduction in the fly's emergence (Razinger et al. 2017).

PERSPECTIVES

Although the potential of various natural enemies have already been studied aiming the control of *D. suzukii*, until this moment, none of the tested species presented significant impact over the pest population as to justify augmentative control. However, since the majority of the aforementioned beneficial species occur naturally in SWD infested crops, they may be able to contribute in the regulation of pest's population regulation. Hence, the use of methods that allow preserving the natural enemies present in a specific area is advised, in order to promote the integration of different management techniques and maintain the population of *D. suzukii* within acceptable levels. In addition, for the success of an integrated management program, it is extremely important to develop and to perfect efficient monitoring methods, as well as establishing control methods in order to determine the precise moment for making a decision.

As to help the natural control performed by the biological control agents present in the production systems, it is advised that conservational methods are preconized, such as using exclusion nets, removing infested fruits, and employing selective insecticides. In addition to it, some techniques have a synergic action in a biological control, like the Sterile Insect Technique, Release of Insects carrying a Dominant Lethal and *Wolbachia*-induced Incompatible Insect Technique, which could complement integrated management control, with no damage to natural enemies. These techniques consist in releasing insects that are sterile (SIT), transgenic (RIDL) or infected with a specific type of *Wolbachia* (IIT) to the nature. When they mate with wild insects those are unable to produce offspring. Currently, different research groups around the world study the integration of such techniques to biological control, and they ought to provide more effective and environmental friendly options for the management of *D. suzukii.*

ACKNOWLEDGMENTS

Authors would like to thank the Coordenação de Aperfeiçoamento de Pessoal de Nível Superior (Capes); Conselho Nacional de Desenvolvimento Científico e Tecnológico (CNPq) for the financial support, and Instituto Nacional dos Hymenoptera Parasitoides (INCT-HYMPAR) for the financial support.

REFERENCES

Allen, W. R. & Hagley, E. A. C. (1990). Epigeal arthropods as predators of mature larvae and pupae of the apple maggot (Diptera: Tephritidae). *Environmental Entomolology*, 19 (2): 309-312.

Ammar, E. D. & Farrag, S. M. (1974). Studies on the behavior and biology of the earwing *Labidura riparia* Pallas (Derm., Labiduridae). *Journal of Applied Entomology*, 75: 189-196.

Anfora, G.; Grassi, A.; Revardi, S.; Graiff, M. (2012). *Drosophila suzukii*: a new invasive species threatening European fruit production. *EnviroChange* 1-7.

Arnó, J.; Riudavets, J.; Gabarra, R. (2012). Survey of host plants and natural enemies of *Drosophila suzukii* in an area of strawberry production in Catalonia (northeast Spain). *IOBC/WPRS Bulletin,* 80: 29–34.

Asplen, M.; Anfora, G.; Biondi, A.; Choi, D-S.; Chu, D.; Daane, K. M.; et al. (2015). Invasion biology of spotted wing drosophila (*Drosophila suzukii*): a global perspective and future priorities. *Journal Pest of Science*, 88: 469–494.

Ben-Dov, E. (2014). *Bacillus thuringiensis* subsp. *israelensis* and Its dipteran-specific toxins. *Toxins*, 6 (4): 1222–1243.

Birken, E. M. & Cloyd, R. A. (2007). Food preference of the rove beetle, *Atheta coriaria* Kraatz (Coleoptera: Staphylinidae) under laboratory conditions. *Insect Science*, 14: 53-56.

Bolda, M. P.; Goodhue R. E.; Zalom, F.G. (2010). Spotted wing drosophila: potential economic impact of a newly established pest. *Giannini Foundation Agricultural Economics*, 13: 5-8.

Brown, P. H.; Shearer, P. W.; Miller, J. C.; Thistlewood, H. M. A. The discovery and rearing of a parasitoid (Hymenoptera: Pteromalidae) associated with Spotted wing drosophila, *Drosophila suzukii*, in Oregon and British Columbia. *ESA Annual Meetings Online Programs*. (2011). 2017/04/09. Available from https://esa.confex.com/esa/2011/webprogram/Paper59733.html.

Bruck, D. J.; Bolda, M.; Tanigoshi, L.; Klick, J.; Kleiber, J.; DeFrancesco, J.; Gerdeman, B.; Spitler, H. (2011). Laboratory and field comparisons of insecticides to reduce infestation of *Drosophila suzukii* in berry crops. *Pest Management Science*, 67: 1357-1385.

Cancino, M. D. G.; Hernandez, A. G.; Cabrera, J. G.; Carrillo, G. M.; Gonzalez, A. S.; Bernal, H. C. A. (2015). Parasitoides de *Drosophila suzukii* (Matsumura) (Diptera: Drosophilidae) en Colima, México. *Southwestern Entomologist*, 40 (4): 855-858. [Cancino, M.D. G.; Hernandez, A. G.; Cabrera, J. G.; Carrillo, G. M.; Gonzalez, A. S.; Bernal, H.C.A. (2015). Parasitoids of *Drosophila suzukii* (Matsumura) (Diptera: Drosophilidae) in Colima, Mexico. *Southwestern Entomologist,* 40 (4): 855-858.]

Calabria, G.; Máca, J.; Bächli, G.; Serra, L.; Pascual, M. (2012). First records of the potential pest species *Drosophila suzukii* (Diptera: Drosophilidae) in Europe. *Journal of Applied Entomology*, 136: 139-147, 2012.

Carrillo, G. M.; Vélz, B. R.; González, A. S.; Bernal, H. C. A. (2015). Trampeo y registro del parasitoide *Pachycrepoideus vindemmiae*

(Rondani) (Hymenoptera: Pteromalidae) sobre *Drosophila suzukii* (Matsumura) (Diptera: Drosophilidae) en Mexico. *Southwestern Entolomogist*, 40 (1): 199-204. [Carrillo, G. M.; Vélz, B. R.; González, A. S.; Bernal, H. C. A. (2015). Trapping and recording of the parasitoid *Pachycrepoideus vindemmiae* (Rondani) (Hymenoptera: Pteromalidae) of *Drosophila suzukii* (Matsumura) (Diptera: Drosophilidae) in Mexico. Southwestern Entolomogist, 40 (1): 199-204.]

Chabert, S.; Allemand, R.; Poyet, M.; Eslin, P.; Gibert, P. (2012). Ability of European parasitoids (Hymenoptera) to control a new invasive Asiatic pest, *Drosophila suzukii*. *Biological Control*, 63: 40-47.

Cini, A.; Ioriatti, C.; Anfora, G. (2012). A review of the invasion of *Drosophila suzukii* in Europe and a draft research agenda for integrated pest management. *Bulletin of Insectology*, 65 (1): 149-160.

Cossentine, J.; Robertson, M.; Xu, D. (2016a) Biological Activity of *Bacillus thuringiensis* in *Drosophila suzukii* (Diptera: Drosophilidae). *Journal of Economic Entomology*, 109 (3): 1071–1078.

Cossentine, J.; Rrobertson, M.; Buitenhuis, R. (2016b). Impact of acquired entomopathogenic fungi on adult *Drosophila suzukii* survival and fecundity. *Biological Control*, 103: 129–137.

Cuch-Arguimbau, N.; Escudero-Colomar, L. A.; Forshage, M.; Pujade-Villar, J. (2013). Identificadas dos espécies de Hymenoptera como probables parasitoides de *Drosophila suzukii* en una plantación ecológica de cerezos en Begues. *Phytoma*, 247: 1-6. [Cuch-Arguimbau, N.; Escudero-Colomar, L. A.; Forshage, M.; Pujade-Villar, J. (2013). Identified two species of Hymenoptera as probably parasitoid of *Drosophila suzukii* in an ecological orchard of cherry in Begues. *Phytoma*, 247: 1-6.

Cuthbertson, A. G. S.; Blackburn, L. F.; Audsley, N. (2014a). Efficacy of commercially available invertebrate predators against *Drosophila suzukii*. *Insects*, 5: 952-960.

Cuthbertson, A. G. S.; Collins, D. A.; Blackburn, L. F.; Audsley, N.; Bell, H. A. (2014b). Preliminary screening of potential control products against *Drosophila suzukii*. *Insects*, 5 (2): 488–498.

Daane, K. M.; Wang, X. G.; Bioni, A.; Miller, B.; Miller, J. C.; et al. (2016). First exploration of parasitoids of *Drosophila suzukii* in South Korea as potential classical biological agents. *Journal of Pest Science*, 89 (3): 823-835.

DAFF (2013). Final pest risk analysis report for *Drosophila suzukii*. Department of Agriculture, Fisheries and Forestry, Australian Government.

De Ros, G.; Anfora G.; Grassi, A.; Ioriatti, C. (2013). The potential economic impact of *Drosophila suzukii* on small fruits production in Trentino (Italy). *IOBC/WPRS Bull*, 93: 317–332.

Deprá, M.; Poppe, J. L.; Schmitz, H. J.; De Toni, D. C.; Valente, V. L. S. (2014). The first records of the invasive pest *Drosophila suzukii* in South American Continent. *Journal of Pest Science*, 87: 379-383.

Dos Santos, L. A.; Mendes, M. F.; Krüger, A. P.; Blauth, M. L.; Gottschalk, M. S.; Garcia, F. R. M. (2017). Global potential distribution of *Drosophila suzukii* (Diptera, Drosophilidae). *Plos One*, 12: e0174318.

Emiljanowicz, L. M.; Ryan, G. D.; Langille, A.; Newman, J. (2014). Development, reproductive output and population growth of the fruit fly pest *Drosophila suzukii* (Diptera: Drosophilidae) on artificial diet. *Journal of Economic Entomology*, 107: 1392–1398.

Englert, C. & Herz, A. (2016). Native predators and parasitoids for biological regulation of *Drosophila suzukii* in Germany. *17th International Conference on Organic Fruit-Growing: Proceedings*, 284-285.

Funderburk, J.; Stavinsky, J.; Olson, S. (2000). Predation of *Frankliniella occidentalis* (Thysanoptera: Thripidae) in field peppers by *Orius insidiosus* (Hemiptera: Anthocoridae). *Environmental Entomology*, 29 (2): 376-382.

Gabarra, R.; Riudavets, J.; Rodríguez, G. A.; Pujade-Villar, J.; Arnó, J. (2015). Prospects for the biological control of *Drosophila suzukii*. *BioControl*, 60: 331-339.

Gao, H.; Xu, N.; Chen, H.; Liu, Q.; Pu, Q.; Qin, Y.; Zhai, Y.; Yu, Y. (2017). Impact of selected fungi from an artificial diet on the growth and development of *Drosophila suzukii* (Diptera: Drosophilidae). *Journal of Asia-Pacific Entomology*, 20 (1): 141-149.

Gargani, E.; Tarchi, F.; Frosinini, R.; Mazza, G.; Simoni, S. (2013). Notes on *Drosophila suzukii* Matsumura (Diptera Drosophilidae): Field survey in Tuscany and laboratory evaluation of organic products. *Redia*, 96: 85–90.

Geisler, F. C. S.; Santos, J.; Holdefer, D. R.; Garcia, F. R. M. (2015). Primeiro registro de *Drosophila suzukii* (Matsumura, 1931) (Diptera: Drosophilidae) para o estado do Paraná, Brasil e de novos hospedeiros. *Revista de Ciências Ambientais*, 9 (2): 125–129. [Geisler, F. C. S.; Santos, J.; Holdefer, D. R.; Garcia, F. R. M. (2015). First record of *Drosophila suzukii* (Matsumura, 1931) (Diptera: Drosophilidae) for the state of Paraná, Brazil and new hosts. *Revista de Ciências Ambientais,,* 9 (2): 125-129.]

Gonzaléz, G.; Mary, A. L.; Scatoni, B.; Lorenzo, M. E.; Vingnale, B.; Goñi, B. (2015). *Drosophila suzukii* (Arthropoda, Insecta, Diptera) en Uruguai. *Resumo...* Congreso Uruguayo de Zoologia. [Gonzalez, G.; Mary, A. L.; Scatoni, B.; Lorenzo, M. E.; Vingnale, B.; Goñi, B. (2015). *Drosophila suzukii* (Arthropoda, Insecta, Diptera) in Uruguay. *Summary* ... Uruguayan Congress of Zoology.

Goodhue, R. E.; Bolda, M.; Farnsworth, D.; Williams, J. C.; Zalom, F. G. (2011). Spotted wing drosophila infestation of California strawberries and raspberries: economic analysis of potential revenue losses and control costs. *Pest Management Science*, 67: 1396-1402.

Guerrieri, E.; Giorgini, M.; Cascone, P.; Carpenito, S.; Achterberg, C. V. (2016). Species diversity in the parasitoid genus *Asobara* (Hymenoptera: Braconidae) from native area of the fruit fly pest *Drosophila suzukii* (Diptera: Drosophilidae). *PLoS ONE*, 11 (2): 1-28.

Hauser, M.; Gaimari, S.; Damus, M. (2009). *Drosophila suzukii* new to North America. *Fly Times*, (43): 12–15.

Hauser, M. (2011). A historic account of the invasion of *Drosophila suzukii* (Matsumura) (Diptera: Drosophilidae) in the continental United States, with remarks on their identification. *Pest Management Science*, 67: 1352–1357.

Haye, T.; Girod, P.; Cuthbertson, A. G. S.; Wang, X. G.; Daane, K. M.; Hoelmer, K. A.; Baroffio, C.; Zhang, J. P.; Desneux, N. (2016). Current SWD IPM tactics and their practical implementation in fruit crops across different regions around the world. *Journal of Pest Science*, 89 (3): 643-651.

Heavner, M. E.; Hudgins, A. D.; Rajwani, R.; Morales, J.; Govind, S. (2014). Harnessing the natural *Drosophila* – parasitoids model for integrating insect immunity with functional venomics. *Current Opinion in Insect Science*, 6: 61-67.

Hodgson, P. J.; Sivinski, J.; Quintera, G.; Aluja, M. (1998). Depth of pupation and survival of Fruit Fly (*Anastrepha* spp.: Tephritidae) pupae in a range of agricultural habitats. *Environmental Entomolology*, 27 (6): 1310-1314.

Jandricic, S.; Scott-Dupree, C. D.; Broadbent, A. B.; Harris, C. R.; Murphy, G. (2006). Compatibility of *Atheta coriaria* with other biological control agents and reduced-risk insecticides used in greenhouse floriculture integrated pest management programs for fungus gnats. *The Canadian Entomologist*, 138: 712–722.

Kacson, B.; Schlenke, T. A. (2012). High hemocyte load is associated with increased resistence against parasitoids in *Drosophila suzukii*, a relative of *D. melanogaster*. *PLoS ONE*, 7(4): 1-16.

Kaneshiro, K. Y. (1983). *Drosophila* (*Sophophora*) *suzukii* (Matsumura). *Proceedings of the Hawaiian Entomological Society*, 24: 179.

Kasuya, N.; Mitsui, H.; Shinsuke, I.; Watada, M.; Kimura, M. T. (2013). Ecological, morphological and molecular studies on *Ganaspis* individuals (Hymenoptera: Figitidae) attacking *Drosophila suzukii* (Diptera: drosophilidae). *Applied Entomology and Zoology*, 48 (1): 87-92.

Keane, R. M. & Crawley, M. J. (2002). Exotic plant invasions and the enemy release hypothesis. *Trends Ecology Evolution*, 17 (6): 164–170.

Kenis, M.; Tonina, L.; Eschen, R.; Van der Sluis, B.; Sancassani, M.; Mori, N.; Haye, T.; Helsen, H. (2016). Non-crop plants used as hosts by *Drosophila suzukii* in Europe. *Journal of Pest Science*, 89 (3): 735-748.

Krzemien, J.; Crozatier, M.; Vicent, A. (2010). Ontogeny of the *Drosophila* larval hematopoietic organ, hemocyte homeostasis and dedicated cellular immune response to parasitism. *The International Journal of Development Biology*, 54: 1117-1125.

Lavine, M. D. & Strand, M. R. (2002). Insect hemocytes and their role in immunity. *Insect Biochemistry and Molecular Biology*, 32: 1295-1309.

Leach, H.; Van Timmeren, S.; Rufus, I. (2016). Exclusion Netting Delays and Reduces *Drosophila suzukii* (Diptera: Drosophilidae) Infestation in Raspberries. *Journal of Economic Entomology,* 109 (5): 2151-2158.

Lee, J. C.; Bruck, D. J.; Curry, H.; Edwards, D.; Haviland, D. R.; Steenwyk, R. A.; Van Yorgey, B. M. (2011). The susceptibility of small fruits and cherries to the spotted-wing drosophila, *Drosophila suzukii*. *Pest Management Science*, 67: 1358–1367.

Lee, J. C.; Dreves, A. J.; Cave, A. M.; Kawai, S.; Isaacs, R.; Miller, J. C.; Van Timmermen S.; Bruck, D. J. (2015). Infestation of wild and ornamental noncrop fruits by *Drosophila suzukii* (Diptera: Drosophilidae). *Annals of the Entomological Society of America*, 108: 117–129.

Miller, B.; Anfora, G.; Buffington, M.; Daane, K. M.; Dalton, D. T.; et al. (2015). Seasonal, occurrence of resident parasitoids associated with *Drosophila suzukii* in two small fruit protection regions of Italy and the USA. *Bulletin of Insectology*, 68 (2): 255-263.

Mitsui, H.; Achterberg, K. V.; Nordland, G.; Kimura, M. T. (2007). Geographical distributions and host associations of larval parasitoids of frugivorous Drosophilidae in Japan. *Journal of Natural History*, 41: 1731-1738.

Mitsui, H. & Kimura, M. T. (2010). Distribution, abundance and host association of two parasitoid species attacking frugivorous drosophilid larvae in central Japan. *European Journal of Entomology*, 107: 535-540.

Monzó, C.; Sabater-Muñoz, B.; Uubaneja, A.; Castañera, P. (2011). The ground beetle *Pseudophonus rufipes* revealed as predator of *Ceratitis capitata* in citrus orchards. *Biological Control*, 56: 17-21.

Naranjo-Lázaro, A.J.M.; Mellín-Rosas, M.A.; Sánchez-González, J.A.; Moreno, G.; Arredondo-Bernal, H.C. (2014). Susceptibility of *Drosophila suzukii* Matsumura (Diptera: Drosophilidae) to Entomophatogenic Fungi. *Southwestern Entomologist*, 39 (1): 201–203.

Nomano, F.Y.; Mitsui, H.; Kimura, M.T. (2015). Capacity of Japanese *Asobara* species (Hymenoptera; Braconidae) to parasitize a fruit pest *Drosophila suzukii* (Diptera; Drosophilidae). *Journal of Applied Entomology*, 139: 105-113.

O'Grady, P.M.; Beardsley, J.W.; Pereira, W.D. (2002). New records for introduced Drosophilidae (Diptera) in Hawaii. *Bishop Museum Occasional Papers*, 69: 34–35.

Peralta-Manzo, J. J.; Lezama--Gutiérrez, R.; Castrejón-Agapito, H.; Mora, J. C.;, Rebolledo-Domínguez, L. A. O uso de *Metarhizium anisopliae* y *Cordyceps bassiana* (Ascomycetes) para el control de *Drosophila suzukii* (Diptera: Drosophilidae) en cultivo de zarzamora (*Rubus fruticosus*). *Entomología Mexicana*, 1: 230–235. [Peralta-Manzo, J. J.;Lezama - Gutiérrez, R.; Castrejón-Agapito, H.; Mora, J. C.; Rebolledo-Domínguez, L. A. Use of *Metarhizium anisopliae* and *Cordyceps bassiana* (Ascomycetes) to control of *Drosophila suzukii* (Diptera: Drosophilidae) in blackberry orchard (*Rubus fruticosus*). *Entomología Mexicana*, 1: 230-235.]

Picchi, M. S.; Marchi, S.; Albertini, A.; Petacchi, R. (2017). Organic management of olive orchards increases the predation rate of overwintering pupae of *Bactrocera oleae* (Diptera: Tephritidae). *Biological Control*, *in press*, 2017.

Pochubay, E.; Touttois, J.; Himmelein, J.; Grieshop, M. (2015). Slow-release sachets of *Neoseiulus cucumeris* predatory mites reduce intraguild predation by *Dalotia coriaria* in greenhouse biological control systems. *Insects*, 6: 489-507.

Powell, J. R. (1997). Progress and prospects in evolutionary biology: the *Drosophila* model. New York: Oxford University Press. 512p.

Poyet, M.; Havard, S.; Precost, G.; Chabrerie, O.; Doury, G.; et al. (2013). Resistence of *Drosophila suzukii* to the larval parasitoids *Leptopilina heterotoma* and *Asobara japonica* is related to haemocyte load. *Physiological Entomology*, 38: 45-53.

Poyet, M.; Le Roux, V.; Gibert, P.; Meirland, A.; Prevost, G.; Eslin, P.; Chabrerie, O. (2015). The wide potential trophic niche of the Asiatic fruit fly *Drosophila suzukii*: the key of its invasion success in temperate Europe? *PloS ONE*, 10 (11): 1–26.

Razinger, J.; Fink, K.; Kerin, A.; Modic, S.; Urek, G. (2017). Susceptibility of Spotted Wing Drosophila (*Drosophila suzukii* (Matsumura, 1931)) pupae to entomopathogenic fungi. *Acta Agriculturae Slovenica*, 109 (1): 125–134.

Renkema, J. M.; Buitenhuis, R.; Hallett, R. H. (2014). Optimizing Trap Design and Trapping Protocols for *Drosophila suzukii* (Diptera: Drosophilidae). *Journal of Economic Entomology*, 107 (6): 2107-2118.

Renkema, J. M.; Telfer, Z.; Gariepy, T.; Hallett, R. H. (2015). *Dalotia coriaria* as a predator of *Drosophila suzukii*: functional responses, reduced fruit infestation and molecular diagnostics. *Biological Control*, 89: 1-10.

Rossi-Stacconi, M. V.; Grassi, A.; Dalton, D. T.; Miller, B.; Ouantar, M.; Loni, A.; Ioriatti, C.; Walton, V. M.; Anfora, G. (2013). First field records of *Pachycrepoideus vindemiae* as a parasitoid of *Drosophila suzukii* in European and Oregon small fruit production areas. *Entomologia*, 1: 11–16.

Rossi-Stacconi, M. V.; Buffington, M.; Daane, K. M.; Dalton, D. T.; Grassi, A., et al. (2015). Host stage preference, efficacy and fecundity of parasitoids attacking *Drosophila suzukii* in newly invades areas. *Biological Control*, 84: 28 -35.

Roy, H. E.; Lawson, H. L. J.; Sschörogge, K.; Poland, R. L.; Purse, B. V. (2011). Can the enemy release hypothesis explain the success of invasive alien predators and parasitoids? *BioControl*, 56: 451–468.

Rutledge, C. E. & O´Neil, R. J. (2005) *Orius insidiosus* (Say) as a predator of the soybean aphid, *Aphis glycines* Matsumura. *Biological Control*, 33: 56-64.

Schlesener, D. C. H.; Nunes, A. M.; Cordeiro, J.; Gottchalk, M. S.; Garcia, F. R. M. (2014). Mosca-da-cereja: Uma nova ameaça a fruticultura brasileira. *Cultivar HF*, 12: 6-8. [Schlesener, D. C. H.; Nunes, A. M.; Cordeiro, J.; Gottchalk, M. S.; Garcia, F. R. M. (2014). Cherry Fly: A new threat to Brazilian fruit growing. *Cultivar HF*, 12: 6-8.

Tochen, S.; Dalton, D. T.; Wiman, N.; Hamm, C.; Shearer, P. W.; Walton, V.M. (2014). Temperature-related development and population parameters for *Drosophila suzukii* (Diptera: Drosophilidae) on cherry and blueberry. *Physiological Entomology*, 43: 1–10.

Urbaneja, A.; Mari, F. G.; Tortosa, D.; Navarro, C.; Vanaclocha, P.; Bargues, L.; Castañera, P. (2006). Influence of ground predators on the survival of the Mediterranean fruit fly pupae, *Ceratitis capitata*, in Spanish citrus orchards. *BioControl*, 51: 611-626.

Walsh, D. B.; Bolda, M. P.; Goodhue, R. E.; Dreves, A. J.; Lee, J.; Bruck, D. J.; Walton, V. M.; O´Neal, S. D.; Zalom, F. G. (2011). *Drosophila suzukii* (Diptera: Drosophilidae): Invasive pest of ripening soft fruit expanding its geographic range and damage potential. *Journal of Integrates Pest Management*, 1–8.

Wang, X. G.; Kaçar, G.; Biondi, A.; Daane, K. M. (2016). Life-history and host preference of *Trichopria drosophilae*, a pupal parasitoid of spotted wing drosophila. *BioControl,* 61 (4): 387-397.

Wang, X. G. & Messing, R. H. (2004). The ectoparasitic pupal parasitoid, *Pachycrepoideus vindemmiae* (Hymenoptera: Pteromalidae); attacks other primary tephritid fruit fly parasitoids: host expansion and potential non-target impact. *Biological Control*, 31: 227-236.

Westphal, M. I.; Browne, M.; Mackinnon, K.; Noble, I. (2008). The link between international trade and the global distribution of invasive alien species. *Biological Invasions*, 10: 391-398.

Wiman, N. G.; Walton, V. M.; Dalton, D. T.; Anfora, G.; Burrack, H.J.; Chiu, J. C.; Daane, K.M.; Grassi, A.; Miller, B.; Tochen, S.; Wang, X.; Ioriatti, C. (2014). Integrating temperature-dependent life table data into a matrix projection model for *Drosophila suzukii* population estimation. *PloS ONE*, 9 (9): 1-8.

Wollmann, J.; Schlesener, D. C. H.; Ferreira, M. S.; Garcia, M. S.; Costa, V. A.; Garcia, F. R. M. (2016). Parasitoids of Drosophilidae with potential for parasitism on *Drosophila suzukii* in Brazil. *Drosophila Information Service*, 99: 38-42.

Woltz, J. M.; Donahue, K. M.; Bruck, D. J.; Lee, J. C. (2015). Efficacy of commercially available predators, nematodes and fungal entomopathogens for augmentative control of *Drosophila suzukii. Journal of Applied Entomology*, 139 (10): 759–770.

In: Biological Control
Editor: Lewis Davenport
ISBN: 978-1-53612-416-3

Chapter 2

BIOLOGICAL CONTROL OF FRUIT FLIES OF THE GENUS *ANASTREPHA* (DIPTERA: TEPHRITIDAE): CURRENT STATUS AND PERSPECTIVES

***Flávio Roberto Mello Garcia*[1,*]*,*
Andressa Lima de Brida*[1]*, Liliane Nachtigall Martins*[1]*,
***Lenon Morales Abeijon*[1] *and Junir Antonio Lutinski*[2]**
[1]Federal University of Pelotas, Pelotas, Brazil
[2]Community University of the Region of Chapecó, Brazil

ABSTRACT

Historically, control of fruit flies has been performed with insecticides applied for full coverage. The use of synthetic insecticides in

* Corresponding author: Department of Ecology, Zoology and Genetics, Insect Ecology Lab, UFPEL, Institute of Biology, Pelotas, RS, Brazil. Email flaviormg@hotmail.com

orcharding has aroused greater care due to market requirements, human and animal health problems and environmental impacts. In this context, the biological control of fruit flies becomes essential, since it aims to control the populations of the pest in balance with the new environmental, health and economic requirements. Biological control of fruit flies can be done mainly with the use of parasitoids, however the use of pathogens (bacteria, fungi and nematodes) and predators can be promising from an Integrated Pest Management perspective. In view of the above, this chapter aims to recover the history and present the current status of the biological control of fruit flies of the genus *Anastrepha* and to show perspectives of the use of such method.

Keywords: control, fruit flies, parasitoid, fungi, bacteria, entomopathogenic nematodes

INTRODUCTION

Approximately 70 species of fruit flies (Diptera: Tephritidae) are considered important agricultural pests (Garcia, 2009). These tephritids are mainly found in fruit crops including oranges, mangoes, apples and peaches, and fall into five genera: *Anastrepha* Schiner, *Bactrocera* Macquart, *Ceratitis* MacLeay, *Dacus* Fabricus and *Rhagoletis* Loew (Garcia, 2009).

According to Morgante (1991), damages caused by fruit flies are classified as follows:

- cultural damages – the damaged fruits are discharged for industrial processing or for *in natura* consumption; infested orchards are depreciated; fruits lose their commercial value; losses can reach up to 100%;
- economic damages – orchards become economically inefficient, leading to investment losses and to the bankruptcy of producers; commercialization of fruits in the external market becomes inviable; foreign currencies surpluses are no longer generated;

fruits from infested areas do not reach profitable prices in internal markets; amount of collectable taxes and duties decline.

- social damages – lay off of workers leads to an increased unemployment and a subsequent migration from rural areas.

The genus *Anastrepha* comprises 212 species, most of them found in Neotropical regions. Yet, some species may be found in the Neoartic Region (Malavasi et al., 2000). Presently, Brazil has the highest number of identified species (Norrbom, 2012), about 47 percent of all species identified in the Americas (Zucchi, 2007). The species considered most deleterious are *Anastrepha ludens* (Mexican fruit fly), *Anastrepha obliqua* (West fruit fly) and *Anastrepha fraterculus* (South American fruit fly). None of them can be considered as an invasive species, for all of them inhabit their probable areas of origin (Malavasi et al., 2000).

Throughout the world, control of fruit flies has been conducted mainly through malathion-based toxic baits (Mangan, 2006). Studies have documented significant impacts of toxic baits on non-target insect species including predators and parasitoids (Bortoli et al., 2016), interfering with biological control (Raga and Sato, 2005) and leaving undesirable residues in fruits (Scoz et al., 2004), which demand the search for management alternatives.

In this context, the biological control of fruit flies emerges as an alternative aiming at the maintenance of agricultural production with low environmental impact and food safety to consumers. For this purpose, pathogens, parasitoids and predators can be used.

The goal of this chapter is to present a review on the main agents of biological control of flies of the genus *Anastrepha*, characterizing whenever possible its viability of use.

BACTERIA

New alternatives for the biological control of pests in agriculture have attracted the attention in the scientific milieu, including the entomopathogenic bacteria, with the purpose of reducing the use of pesticides, acting in the control of insects causing damage (Ferry et al., 2004). *Bacillus thuringiensis* (Berliner) (*Bt*) is a microbial agent used in pest control in agriculture, and is considered a safe method for humans and the environment (Campanini et al., 2012).

B. thuringiensis is a gram-positive rod, has vegetative cells and is mobile in the environment. It is naturally found in soil, but also in the substrate of plants, on surfaces and in tissues of plants, in addition to stored products and in insects. It grows by aerobiosis, but also possesses facultative anaerobiosis (Alves, 1998; Polanczyk et al., 2008).

Within the genus *Bacillus* all species produce endospores (*B. thuringiensis, B. cereus, B. anthracis, B. moycoides* and *B. weihenstephanensis*). This group is responsible for more than 90% of biopesticides available worldwide (Polanczyk; Alves, 2003). The main characteristic that distinguishes *B. thuringiensis* from other taxa of the same genus is the presence of a protein crystal (Vilas-Bôas et al., 2007). These crystals have entomopathogenic activity, being characterized as bioinsecticides mainly against insects belonging to the orders Lepidoptera, Diptera, Hymenoptera and Coleoptera (Schnepf et al., 1998).

The crystals of *B. thuringiensis*, also known as Cry proteins, are formed by crystal-shaped proteins, the δ-endotoxins (Bravo et al., 2007). Cry proteins are synthesized as protoxins, their action depends on the activation processes that occur inside the digestive tract of the insect (Chesthukina et al., 1982). After ingestion, crystals are solubilized in the insect gut (alkaline pH), releasing the protoxins that are cleaved by proteases of the insect itself, resulting in active toxins. These toxins are able to bind to specific receptors present in microvilli of the insect's intestinal cells (Bravo et al., 2007). The toxins with specific receptors lead to the formation of oligomers, which bind to secondary receptors in the intestinal cell membrane. As a result of this binding, the oligomeric toxin is inserted into the membrane of the intestinal epithelial cell, resulting in pores in this epithelium (Alves, 1998). The action of toxins results in paralysis of the digestive tract, leading to death by starvation, general paralysis of muscles and septicemia (Vallete-Gelly et al., 2008).

The Mexican fruit fly is a threat to citrus cultivation in Central and North Americas (Robacker et al., 1996). Thinking of a technique less harmful to the environment and that could be applied to the management of *Anastrepha ludens* (Loew). Robacker et al., (1996) investigated the toxicity of 55 *B. thuringiensis* isolates in *A. ludens* larvae, of which only seven were toxic for Mexican fruit fly larvae and five for adults, which reached a mortality of 65-80% adults fed for 10 days.

In this first study on the genus *Anastrepha* together with the entomopathogenic bacterium *B. thuringiensis*, Robacker et al., (1996) collected samples of soil and dead flies to isolate these bacteria and test their entomopathogenic potential against *A. ludens*. They showed that the different isolates had different pathogenic action for each stage of the fly cycle and demonstrated that the most toxic isolate for larvae was not the most toxic for adults, therefore indicating the need for prior identification of the material that is intended to use, characterizing which phase of the cycle to be controlled and thereby separating the appropriate isolate.

In recent years, this technique for collecting soil material to verify the presence of *B. thuringiensis* has been increasingly common in order to obtain environmental samples, which contain non-toxic strains (Ohba; Aizawa, 1986; Hofte; Whitely, 1989). When collecting soil samples, the proportion of different strains of *B. thuringiensis* suggests that the strains are better conditioned and therefore become more protected in the soil, at a depth of about 5cm, in these substrates, *Bt* is protected from the sunlight and also from unfavorable temperatures (Bel et al., 1997).

In the study by Robacker et al., (1996), the adult mortality rate was around 80%, four days for insect mortality, may be an aggravating factor when thinking about a technique that aims to use microbial agents for control in toxic bait. Therefore, the 20% insects that do not die can acquire resistance if management is not suitable and spread this characteristic to the next population (Starnes et al., 1993).

In order to use this type of entomopathogen in the control of insects it is necessary to force their feeding, because it is a microbial agent that necessarily needs to be ingested and after this ingestion, the insect dies. Knowing the behavior of hosts in question, fruit flies, which requires for their development energy source like carbohydrates and proteins, the researchers that study this form of control use culture media as the artificial diets to feed the flies. Martinez et al., (1997) used granulated sugar and a type of fermented associated with *B. thuringiensis* var. *dannstadiensis* and offered to adults of *A. ludens*.

When checking the pathogenic action of this variety on adult feeding, which lasted 4 minutes, Martinez et al., (1997) found a mortality of 70% and in 30 minutes, the observed mortality reached 90%, demonstrating this toxicity to insects from one to 21 days of age, aiming to replace the insecticide used in toxic bait with this variety of *Bt*.

Timed mortality in *Bt*-based bioassays indicates that as feed time increases, the flies will have more access to the bacteria and thus a greater probability of ingesting *Bt* spores and dying. For this, it is necessary that the commercial products containing *B. thuringiensis* have a central focus on the control of adults, since the larvae are inserted inside the fruits, and the adults have free life, therefore, a variety with capacity to cause mortality to adults will reduce the population of the pest and consequently the damage caused to the fruit trees (Martinez et al., 1997).

B. thuringiensis can express different types of Cry toxins, however not all are viable to be used in biological control. When analyzing the toxic effect of *B. thuringiensis* b-exotoxin on three species of *Anastrepha* (*A. ludens*, *Anastrepha obliqua* (Macquart) and *A. serpentina* (Wiedemann), it can be observed that b-exotoxin was highly toxic to all three species tested. As a consequence of this effect, there were deformations in pupae, as well as reduction in longevity, fecundity and egg hatching (Toledo et al., 1999).

Toledo et al., (1999) demonstrated that the deleterious effect of b-exotoxin was relatively low to larvae, however, the pupae presented patterns of deformations that influenced the emergence of adults. The use of b-exotoxins should be avoided because they have a toxic action in vertebrates, which could lead to contamination of the applicators in the field. B-exotoxin is not registered for use in the United States; it interferes with the synthesis of RNA in vertebrates and invertebrates, considered an environmental hazard by the Environmental Protection Agency (Mc-Clintock et al., 1995).

The toxic principles of b-exotoxin are the target of several studies, through which it can be observed high mortality in fruit flies (Martinez et al., 1997; Toledo et al., 1999). Nevertheless, Robacker et al., (1996) demonstrated that b-exotoxin alone is not toxic to adults of *A. ludens*, what is actually toxic are endotoxins/spores in powder form.

Buentello-Wong et al., (2015) developed a large and diverse collection of *B. thuringiensis* with potential against different agricultural pests, some of which have an action against *A. ludens*. Using technical combinations when isolating the samples of *B. thuringiensis*, which can determine toxic action, aid in the global distribution and reduce the environmental impact contributing to the control of dipterans.

The entomopathogen *B. thuringiensis* is an alternative tool that should be better explored and studied for application in the integrated management of fruit flies, since *Bt*-based products have not yet been developed to control tephthites. The works involving the genus *Anastrepha* are directed to the laboratory and the semi-field. Several varieties of *Bt* are still in the test phase aiming at a strain 100% efficient against target insects in a second way to humans and the environment.

FUNGI

Fungi represent one of the most numerous and heterogeneous groups of living beings found in the most diverse ecological niches. They are responsible for the production of organic acids, enzymes of industrial interest and for the biological control of insects with pest status in agriculture, acting as an entomopathogenic agent helping to reduce insect populations (Espósito; Azevedo, 2004).

Approximately 80% of insect diseases in the field are related to etiological agents of entomopathogenic fungi, the fungi involved belong to 90 genera and more than 700 species (Alves, 1998). Entomopathogenic fungi (EFs) have a potential applicability to the control of agricultural pests (Ffrench-Constant, 2005), due to the high genetic variability they have because they are highly virulent microbial agents, with adequate characteristics to be used as microbial insecticides (Alves, 1998).

With the application of EFs in the field, it is possible to avoid the application of chemicals, reducing environmental contamination. This technique of environmental maintenance represents an ecological advance for the management of agricultural pests. The use of EFs, in addition to reducing environmental damage caused by pesticides, are capable of attacking several species of insects, have low toxicity to other organisms, can be specific to the host (infecting different stages during development), and the infection by fungi is multifactorial, causing insects to have a lower probability of acquired resistance (Alves, 1998).

Epizootic observations of the pathogenic action of natural EFs encourage intense studies to apply them to biological control (Feng et al., 1994). The attack to arthropods occurs after the external contact of the mycelium of the fungus with the cuticle of the host. Soon the process of infection begins with the adhesion of the conidia (vegetative form of the fungus) with the first layer of epicuticle. The adhesion mechanism involves different hydrophobic, electrostatic and physiological interactions. When adhered, the conidium forms a germ tube filled with hyphae, which penetrate into the cuticle of the host. Hyphae branch out reaching the hemocele, toxins from the fungus lead the host to death. After the death of the insect, the fungus continues to develop, invading digestive, nervous and muscular systems. In the spiracles and on the cuticle the hyphae emerge and the sporulation occurs. In the form of conidia the fungus is disseminated again in the environment, by water, wind and other animals, to the point of infecting other hosts and ensure species perpetuation (Alves, 1998; Meyling; Eilenberg, 2007).

Among the most commonly, used EFs in biological pest control are *Beauveria bassiana* (Vuillemin), *Metarhizium anisopliae* (Metsch), *Paecilomyces fumosoroseus* (Wise). They are fungi with wide geographic distribution, easily found hosting insects under natural conditions, causing enzooty and epizooty (Alves, 1998). However, the use of EFs has become common for some species of *Anastrepha*, research is restricted to tests on larvae, pre-pupae, pupae and adults, in laboratories or semi-field.

Species of fruit flies use the soil to reach the pupal stage, this medium when being exploited, is ideal for the release of biological control agents that aim at the elimination of 3rd instar larvae and pupae whose survival depends on favorable environments and without the influence of natural enemies capable of negatively influencing their development (Lezama-Gutiérrez et al., 2000; Ekesi et al., 2003).

B. bassiana EF has cosmopolitan distribution, which makes it more applied and tested in pest control. Causing epizooty, it is characterized by high growth rate, high production of infectious units, ability to survive in the environment and ease of penetration through the integument, in addition to reaching the hemolymph of the host, reaffirming its high pathogenicity (Fuxa, 1987). Because it presents an excellent potential as a biocontrol agent of different dipterans, it has been used in the biological control of tephthids.

Anastrepha ludens (Loew) is one of the most important species in the family, infesting crops such as mango and citrus (Aluja, 1994). Several studies were carried out using *B. bassiana* in the control of *A. ludens* to complement the management of this pest. However, most of the control measures of this species still depend on the use of insecticides that have detrimental effects on invertebrates and non-target insects, besides social and environmental problems (Quesada-Moraga et al., 2008; Dimbi; Maniania, 2009; Toledo et al., 2010).

Evaluating the effect of eight strains of *B. bassiana* on larvae, pupae and adult females of *A. ludens*, Rosa et al., (2002) identified a low mortality in the immature stages (larvae and pupae), identifying a minimal pathogen response in relation to the larvae, causing mortality of 2 to 8%. This effect is due to the extensive sclerosis of the *A. ludens* larvae, which have spikes in the cephalic region, which may have prevented fungus penetration into the cuticle (Aluja, 1994). In other cases, the cuticle was also the great ally of insects to provide protection against fungal mycelia, being able to change its density and thickness, as well as the degree of sclerotization (Hajek; Leger, 1994). Possibly, larvae of *A. ludens* contain substances that inhibit the development of fungi, such as caprylic acid. In relation to the pupae of *A. ludens*, none of the strains of *B. bassiana* were pathogenic, so the fungal mycelium, there was no development, the pupae of this fly have a hard and scaly cuticle, which possibly reduced the contact of the fungus (Rosa et al., 2002).

For adults, the different strains of *B. bassiana* were pathogenic, resulting in 100% mortality. The contact with the fungal mycelium resulted in sporulation through the soft parts of the exoskeleton, which can be identified in the bases of the wings, mouth, intersegmental regions of the legs, the abdomen, coxa and neck. All infected individuals presented a dry and hard appearance, with a pink or reddish coloration, demonstrating that *B. bassiana* is a good candidate to be used to control adults of *A. ludens* (Liedo et al., 2002). Variations in mortality depended on the strain used, its origin and mode of infection; this response will only be possible through pathogenicity tests (Hall, 1982).

Sterile males of *A. ludens* when infected with *B. bassiana* were able to infect females during copula in mating attempts. The females were not infected, the males besides having undergone a sterilization process, were contaminated with fungal mycelia. This process resulted in 65% mortality with *B. bassiana* and 55% with *M. anisopliae* (Toledo et al., 2007).

The viability of horizontal transmission by combining sterility techniques with entomopathogens is a new opportunity for integrated pest management programs, with a concept of self-dissemination. Sterile insects, in addition to reducing pest populations, also act as conidia transmitters, drastically reducing the fertility and longevity of females, which in their entirety are harmful to crops, causing damage to fruits and economic losses. The sum of factors is related to an additional advantage of using these techniques to control *A. ludens* (Toledo et al., 2007).

The competitiveness of sterile and non-sterile *A. ludens* males contaminated with *B. bassiana* is not different, indicating that the presence of the fungus does not reduce mating performance. The difference found suggests that the sterile males are less susceptible to the presence of the fungus compared to the wild males, besides that they have a great advantage, since they are sterilized and do not have the capacity to perpetuate the species, supporting the idea that the sterile males act as vectors of *B. bassiana*, and at the same time reduce the exposure and harmful effects on target organisms (Novelo-Rincón et al., 2009).

As for the virulence of native strains of *B. bassiana* in relation to mortality, ovarian development, oogenesis, survival, fecundity and fertility of *A. ludens*, at different ages, it can be observed that the average mortality is 90%. In this study it is notorious that ovarian development was delayed, so there was no reduction in ovary size. The distribution patterns of vitellogenic follicles were similar between females contaminated with fungi and those not contaminated. Meantime, survival rapidly decreased 5 days after inoculation of the fungus, as well as egg laying and total fecundity, the fertility of the flies was significantly affected (Sánchez-Roblero et al., 2012).

The native strain of *B. bassiana* affected *A. ludens* females, but its biological activity was moderate compared to other strains. The effect of infection on fecundity was more drastic in younger females treated when the oogenesis process was beginning. In flies treated at 8 days, the effect on net fecundity was reduced, probably because the eggs had already been produced (Quesada-Moraga et al., 2006).

The first report on the dynamics of oogenesis in *B. bassiana*-treated *A. ludens* females was observed by Toledo et al., (2010). Although *B. bassiana* has caused low mortality in *A. ludens*, reducing the fecundity of infected flies may reduce its potential for fruit infestation, which is beneficial in reducing the population of this pest (Toledo et al., 2010).

Changes in fruit fly ovary development are associated with fecundity, according to most reports when observing that *B. bassiana*-treated females presented lower fecundity, but their ovarian development was not affected (Perez et al., 2000). Despite the decrease in fecundity, there was no reduction in the size or appearance of the ovaries after infection. This is because the infection did not prevent the production of oocytes, which accumulated in the ovaries, filling the spaces that remained in the basal, secondary areas and tertiary oocytes that did not reach their maximum size. The late vitelogenesis indicates that there was a reduction in supply required for oocyte maturation as a result of infection and colonization of the insect hemolymph by fungal infection, which has been reported for other insect hosts (Hajek; Leger, 1994).

Another possible explanation for the differences between the pathogenicity of the different varieties of native fungi could be associated with inoculation methods. Flies can be contaminated by spraying or dipping in suspension containing conidia or dry conidia. By both methods, the flies acquire a greater amount of inoculum, increasing the mortality and reduction of the LC50.

M. anisopliae is found inhabiting different types of soils. It is categorized as a green fungus due to the greenish color of the colonies in sporulation. It is found naturally on more than 300 species of insects of different orders (Alves, 1998). It penetrates the insects through the cuticle, and once inside the insect, the fungus produces a lateral extension of hyphae that eventually proliferate and consume the tissues of the internal organs. The hyphae growth continues until the insect is fully colonized by mycelia (Cole, 2003). Studies have indicated that *M. anisopliae* is unable to grow at temperatures above 35°C. The results of the toxicity/pathogenicity studies showed no toxic, pathogenic or adverse effect. Although *M. anisopliae* is not infectious or toxic to mammals, inhalation of spores can cause allergic reactions in susceptible individuals. Therefore, it is the least used as a pest control agent, due to the fact that it can become toxic to the applicator (Alves, 1998).

Using *M. anisopliae* against larvae, pre-pupae, pupae and adults of *Anastrepha fraterculus* (South American fruit fly) responsible for losses in temperate fruits (Garcia; Norrbom, 2011), it can be observed that in the larval, pre-pupal, pupal, and adult stages, the fungus infection occurred from the larval phase, but it was present in the pupae and dead adults. The fungus was incorporated in the soil due to the behavior of the larvae, which when reach the 3rd instar go to the substrate to pupate, in this case the calculated mortality was 86% (Destéfano et al., 2005).

P. fumosoroseus as well as other EFs is commonly found in soil, and is also isolated from a variety of insects worldwide (Canton; Vandenberg, 1998). Because of its ability to promote epizooty in different regions of the world, it begins to be explored and developed as a possible biocontrol agent (Jackson et al., 1997). Carneiro and Salles (1994) verified that the CG 260 isolate of *P. fumosoroseus* caused 100% mortality of pupae of *A. fraterculus* after 20 days of application on the pre-pupae larvae, and the few infected larvae had initial symptoms of colonization, with the mycelium expanding from the ends to the whole body. In pupae, they observed lesions or mycelium coming out of the natural openings, a yellowish liquid (5 to 10 days later).

Only the application of conidia to the soil to control fruit flies was published Dimbi et al., 2003; Lezama-Gutierrez et al., 2000). The soil is shown to be an excellent substrate, containing the pupae of the flies to be infected and consequently the newly emerged adults (Ekesi et al., 2005), besides the soil provides conditions necessary for the development of fungi, as well as moisture and ideal temperature.

The social and environmental problems associated with the use of insecticides for the control of fruit flies are the targets of major debates around the world. New alternatives for the biological control of these insects, including entomopathogenic agents, such as fungi are promising and have been studied and discussed by several authors.

Entomopathogenic Nematodes

Among the control alternatives, entomopathogenic nematodes (EPNs) have been highlighed due to their innumerable advantages, from the rapid mortality of the host insect to the method of application of these agents (Fuga et al., 2012).

Nematodes are non-segmented cylindrical worms, belong the Nematode Phylum and are among the most numerous organisms on the planet. The order Rhabditida comprises the two families Steinernematidae and Heterorhabditidae, which are frequently studied in the biological control of pests, which are associated with symbiotic bacteria of the genus *Xenorhabdus* (Thomas; Poinar) e *Photorhabdus* (Boemare, Louis and Kuhl), respectively (Poinar; Grewal, 2012; Batalla-Carrerra et al., 2016). In the soil, these nematodes are found in the third infective juvenile stage (J3), which is responsible for the search and infection of the host. In this stage, the nematode does not feed, and can resist weathering for a long time (Glazer, 2002). The third-stage juvenile (infectious) (IJ) carries its mutualistic bacteria in the intestine, which are highly pathogenic to insects (Poinar; Grewal, 2012). By penetrating a host through natural openings such as the anus, mouth and spiracles or via the integument and invading its hemocele, the nematodes release the bacterium, causing septicemia and leading the insect to death between 24 and 48 hours (Ferraz, 1998; Lewis et al., 2006). EPNs are distributed globally, with different species and groups according to geographic regions (Tumialis et al., 2016).

Entomopathogenic nematodes have been captured on all continents (except Antarctica) and on some islands, and have been used in North America, Europe, Asia and Australia for the control of soil pests and cryptic environments (Grewal et al., 2001), and some species are considered cosmopolitan such as *Steinernema carpocapsae*, *Steinernema feltiae* (Filipjev, 1934), *Heterorhabditis bacteriophora* (Poinar, 1976) and *Heterorhabditis indica* (Poinar et al., 1992) (Barbosa-Negrisoli et al., 2009).

Although information on EPNs and their bacteria is scarce in many countries, including Brazil, the search for new isolates with potential for pest control is growing every year, and some species such as *Heterorhabditis amazonensis* Andaló and *Steinernema brasiliensis* Nguyen have been reported as native to Brazil (Andaló et al., 2006; Nguyen et al., 2010) and *H. indica* and *Heterorhabditis baujardi* were reported in Rondônia, Brazil (Dolinski et al., 2008), *H. amazonensis, H. baujardi, H. indica* and *Heterorhabditis mexicana* were found in Minas Gerais, Brazil (Andaló et al., 2006; Nguyen et al., 2006) and *H. amazonensis, Steinernema rarum* in areas of agricultural and native vegetation in different Brazilian regions (Brida et al., 2017). The first record of EPNs in dipterans was carried out by Poinar (1975), who found *Steinernema* IJs in pupae of *Rhagoletis pomonella* (Walsh) (Diptera: Tephritidae), since then, studies have been carried out to investigate efficiency, virulence and dose of control of different EPNs isolates to combat distinct fruit flies of economic importance.

For the genus *Anastrepha*, investigations have confirmed the virulence of different EPNs isolates as agents for potential control in the laboratory and in the field; among species of fruit fly susceptible to NEPs are *Anastrepha ludens* (Loew), *A. obliqua* (Macquart), *A. serpentina* (Wied.) and *A. fraterculus* (Wied.) (Toledo et al., 2005a, 2005b, 2006a, 2006b; Barbosa-Negrisoli et al., 2009) and researches show results with different mortaility rates, mainly at the larval stage of flies of the genus *Anastrepha*. Toledo et al., (2005a) evaluated the efficiency of *H. bacteriophora* against larvae of *A. ludens* (Loew), obtaining a LC50 of approximately 115 IJ/cm^2 in the laboratory. The effectiveness of *S. carpocapsae* and *S. riobr*ave were confirmed under field conditions, both isolates reduced the emergence of *A. ludens* adults by 64 to 14% (Lezama-Gutiérrez et al., 2006). Under field conditions, mortality of 47 and 76% in concentrations of 115 and 345 IJ/cm^2, respectively, was verified; this same nematode caused mortality to third instar larvae of *A. serpentina* after exposed to a concentration of 36.0 ± 5.4 IJ/cm^2 (Toledo et al., 2006b).

H. bacteriophora JIs had the highest infection and mortality rates in *A. obliqua* larvae when the temperature ranged from 24 ± 0.5°C and the lowest mortality rate was found at lower temperatures 18 ± 0.5°C (Toledo et al., 2005b). *S. carpocapsae, S. riobraves* and caused to 90% mortality to third instar larvae of *A. ludens* and *H. bacteriophora* NC, *S. feltiae* and *S. carpocapsae* Tecoman, 83, 81 and 76% respectively (Lezama-Gutiérriz; Molina, 1996).

The stage of development of the fly is considered important due to its susceptibility to the penetration of the nematoide. Several biological stages of *A. suspensa* were tested using different isolates of *Heterorhabditis* and *Steinernema*, and mortality of larvae and adults were up to 90.7 and 91.7%, respectively (Beavers; Calkins, 1984).

The highest infection rate of *H. bacteriophora, H. indica, S. feltiae, S. carpocapsae* in *A. suspensa* were observed at the larval stage, in contrast, the pupae were resistant to nematode penetration (Heve et al., 2016). *A. ludens* pupae of 5 to 12 days old were less susceptible to infection by

Steinernema feltiae when compared to third instar larvae (Toledo et al., 2001). Kamali et al., (2013) suggest that the main reason for the low infection rate and mortality of *Anastrepha* pupae may be related to the hard integument of the last instar larvae, thus preventing penetration through the pupario. However, pupae of *A. fraterculus* were susceptible to *H. bacteriophora* and *S. riobrave* (Barbosa-Negrisoli et al., 2009), these same nematodes caused mortality of 30 and 21% in pupae of *A. ludens* (Hernández, 2003).

A. fraterculus has been studied with the use of different species of entomopathogenic nematodes and has shown satisfactory responses, with high mortality rates. Of the 20 EPNs isolates tested in *A. fraterculus*, the nematodes *S. carpocapsae* and *Oscheios* sp. were considered pathogenic to this insect (Foelkel et al., 2017). Rodrigues-Trentini (1996) confirmed the high virulence of *S. carpocapsae* to *A. fraterculus* larvae. *H. amazonensis* CB 24 shows the highest virulence to *A. fraterculus,* controlling 70% of the fly population in the laboratory and 67% in the field, in a citrus orchard (Canesin, 2011). *S. carpocapsae* caused the highest percentage of mortality in *A. fraterculus*, this nematode is able to kill from 50 to 90% of the larvae population with approximately 96.3 and 314.7 IJs, respectively (Foelkel et al., 2016). *H. bacteriophora* RS88 and *S. riobrave* RS59 resulted in the highest larval mortality of *A. fraterculus* at pre-pupae and pupae stages, with 1630, 457 and 2851, 423 IJs/cm^2, respectively, and in the field, larval mortality was 51.3%, 28.1% and 20%, 24.3%, respectively (Barbosa-Negrisoli et al., 2009).

Regarding the results obtained by the authors mentioned above, we can draw a sketch that still needs to be researched, so future perspective with the use of entomopathogenic nematodes are based on the search for new isolates with a high level of infectivity for fruit flies. Studies conducted mainly in the field should aim at intensifying the potential of control of these agents, since most species of the genus *Anastrepha* are susceptible to several species of EPNs, mainly at the larval stage. Thus, the use of EPNs in the control of flies becomes an additional tool for biological control to be improved in the practice of integrated pest management, contributing to strategies of handling of fruit flies in the world's fruit farming.

PREDATORS

The importance of parasitoids as agents for the biological control of fruit flies has inspired several studies, but the role of predators has attracted relatively little attention and their contribution to population regulation is unclear (Sivinski, 1996).

Several authors have been studying beneficial arthropods (predators) that inhabit the most diverse agricultural ecosystems and complement that predatory insects of larvae and pupae of fruit flies, registered for the species of *A. fraterculus*, *A. ludens* and *A. suspensa*, belong to the families Carabidae, Formicidae, Labiduridae, Mutility and Staphylinidae (Table 1).

The activity of predatory insects of *A. fraterculus* was reported by Galli and Rampazo (1996), presenting eight species preying upon larvae and pupae, three of which belong to the family Carabidae, two to Staphylinidae and Formicidae, one species of Labiduridae and a morphospecies of Mutillidae in an experiment in the Monte Alto region, São Paulo, Brazil, in guava (*Psidium guava*) with five-year-old trees of the cultivars Paluma and Rica.

In the same study, the above mentioned authors collected in greater intensity the species *Calleida* sp., *Scarites* sp. and *C. granulatum*, but these were not statistically significant.

The family Carabidae presents mostly polyphagous predatory species although there are those that have high specialization in feeding habits (Ball; Bousquet, 2000). Predatory species feed mainly on mollusks, oligochaetes, spiders, and other small arthropods, mainly insects, including their eggs and pupae (Lovei; Sunderland, 1996; Kromp, 1999; Holland; Luff, 2000; Yamazaki; Sugiura, 2006).

The records of staphylinids, dermapterans and others predators of *Anastrepha* are extremely scarce, only occurring in one study (see Galli; Rampazo, 1996). Probably the incipience of the knowledge of predatory activity of rove beetles, such as those collected in the aforementioned study *B. herrhoidalis*, *B. ruffipenis* and *X. analis*, is due to their feeding habit, which may be detritivore or carnivore (Lima et al., 2010), being the latter composed by generalist predatory species or specialists at the larval and adult stages (Rafael et al., 2012). In relation to dermapterans, the genus *Labidura* (Dermaptera: Labidurida) and the species *Euborellia annulipes* (Dermaptera: Anisolabididae) are linked to insect-pest management programs of different crops and are predators of eggs and young forms of Lepidoptera, Hemiptera, Coleoptera and Diptera (Costa et al., 2007; Silva et al., 2010).

Table 1. List of predators of *Anastrepha* with indication of the country where the study was conducted

Fruit flies	Predator	Country	Reference
A. fraterculus	*Calleida* sp.	Brazil	Galli; Rampazo (1996)
	Calosoma granulatum	Brazil	Galli; Rampazo (1996)
	Belonuchus herrhoidalis	Brazil	Galli; Rampazo (1996)
	Belonuchus rufipennis	Brazil	Galli; Rampazo (1996)
	Labidura sp.	Brazil	Galli; Rampazo (1996)
	Pheidole sp.	Brazil	Galli; Rampazo (1996); Fernandes et al., (2012)
	Scarites sp.	Brazil	Galli; Rampazo (1996)
	Solenopsis sp.	Brazil	Galli; Rampazo (1996)
	Mutillidae sp.	Brazil	Galli; Rampazo (1996)
A. ludens	*Belonuchus rufipennis*	?	Aluja, 1994; Stibick, 2004
	Solenopsis geminate	USA Mexico	Stibick (2004); Thomas (1995)
	Xenopygus analis	Mexico	Stibick (2004); Baker et al., (1944)
A. suspensa	*Solenopsis invicta*	Mexico	Stibick (2004); Hennessey (1997)
	Euborellia annulipes	USA	Stibick (2004); Hennessey (1997)

Parasitoid wasps of the family Mutillidae, popularly known as velvet-ants, have as host several orders of insects such as most hymenopterans, coleopterans, dipterans, cockroaches (Blattodea), lepidopterans and even hemipterans (Brothers et al., 2000). According to Townsend et al., (2006) and Ricklefs (2010), parasitoidism can be classified as a form of predation that acts in the process of population control and dynamics, being an ecological correspondent of importance in the ecosystem.

Ants are an efficient group of predatory insects that regulate insect populations in general (Eskafi; Kolbe, 1990; Radeghieri, 2004) and can be considered as pest control agents in agroecosystems. The predation of fruit flies by ants occurs when the larva leaves the fruit to bury itself in the soil for pupation (Fernandes et al., 2012). This larvae accommodation is strongly influenced by soil physical properties, causing the larva to take more time to bury themselves in dry soils, extending the time they remain exposed to predation by a variety of predators (Aluja et al., 2005).

Ants of the genus *Pheidole* were observed as the most important predatory species of *Anastrepha* spp larvae. reaching up to 93% predation on larvae of *Anastrepha* spp., and the species *Pheidole oxyops* Forel, 1908 responsible for the removal of 67.4% larvae offered in an experiment conducted in orchards of guava (*P. guava*), jaboticaba (*Myrciaria jaboticaba*) and mango (*Mangifera indica*) of the Federal University of Grande Dourados (UFGD) (State of Mato Grosso do Sul, Brazil, 22º13'16"S;54º48'20"W) (Fernandes et al., 2012).

In Mexico, in citrus orchards in the city of General Terán, Nuevo Leon, the ant *Solenopsis geminata* (Fabricius, 1804) was responsible for the predation of 95% of *A. ludens* larvae (Thomas, 1995).

The predatory and defensive strategies of organisms are among the most discussed topics in ecology and evolution (Thompson, 1994). Thus, in the case of fruit flies, especially those belonging to the genus *Anastrepha*, rapid penetration into the soil is the best strategy to prevent predation (Thomas, 1995; Aluja, 2005), in which soil characteristics and ability of the larva to bury are determinant for its survival (Fernandes et al., 2012). The lack of moisture in the soil can lead to mortality of pupae newly emerged and adults, who have difficulty to cross dry soils (Baker et al., 1944) and significantly affect the predation rate by ants (Fernandes et al., 2012).

However, studies on insects preying upon *Anastrepha* are still scarce. Furthermore, it is necessary to understand the role that these species can play in conservative biological control in order to provide better environmental conditions in agroecosystems, making them an integral part of integrated pest management programs.

Parasitoids

Parasitoids can be classified according to the way they exploit the host in two categories (Askew; Shaw, 1986):

- ***Koinobionts*** – Parasitoids that allow the host to grow and continue to feed after parasitism. Thus, for example, all the representatives of Figitidae are koinobiont endoparasitoids of Diptera larvae, emerging from the puparium.
- ***Idiobionts*** - Parasitoids that kill their hosts prior to emergence and thus develop in dead or paralyzed hosts. The idiobiont parasitoids of *Anastrepha* pupae attack their hosts when they are in soil, represented by five species of the genus *Coptera* and *Trichopria* (Diapriidae) and three polyphagic species, *Pachycrepoideus vindemiae* (Rondani), *Spalangia cameroni* Perkins e *S. e*ndius Walker (Pteromalidae) (Ovruski et al., 2000).

Parasitoids of *Anastrepha* species belong to five families: Braconidade, Diapriidade, Eulophidae, Figitidae and Pteromalidae, most of which belong to the first family (Ovruski et al., 2000).

Braconidae

The subfamilies of Braconidae important in the biological control of fruit flies are Opiinae and Alysinae, such insects are koinobiont parasitoids, that is the egg laying occurs inside the eggs and larvae, remaining until the pupal stage, when there is the emergence of the adult (Wharton et al., 1997).

Diaschamimorpha longicaudata is an Indo-Asian parasitoid of *Bactrocera* larvae; however, after its introduction in several regions, it started to parasitize species of *Anastrepha* and *Ceratitis capitata*. In addition, this species is one of the most significant biological control agents for augmentative biological control of fruit flies in Latin America (González et al., 2007).

Specimens of *D. longicaudata* that develop in *A. fraterculus* have a higher net reproduction rate (R_0) and produce larger individuals. The daily superparasitism in *A. fraterculus* larvae (1.6 ± 0.22) is higher than in *C. capitata* (0.4 ± 0.07). In suitable environments, *D. longicaudata* can be established in populations of *A. fraterculus*, as it is possible to achieve a successful mass rearing of the parasitoid in this host (Meirelles et al., 2013).

Studies carried out in greenhouse proved that *D. longicaudata* females were able to recognize and parasitize *A. fraterculus* in both native and exotic fruits. However, the number of offspring produced daily is lower than that observed in laboratory (Meirelles et al., 2016)

In this context, parasitism can be complemented by species such as *Fopius arisanus* (Sonan) (Hymenoptera: Braconidae), an egg parasitoid and *Coptera haywardi* (Oglobin) (Hymenoptera: Diiapridae), a pupal parasitoid of fruit flies (Cancino; Montoya, 2006).

Doryctobracon areolatus is a koinobiont parasitoid of 2nd instar larvae of *Anastrepha*. It is the most widely distributed Neotropical/subtropical parasitoid, from southern South America to Florida in the United States (Sivinski et al., 1997).

In Brazil, rearing of *D. areolatus* is done in cages of varying sizes and always uses *A. fraterculus* as the host. The diet for this parasitoid is composed of honey, distilled water, agar, ascorbic acid, and sodium benzoate or honey and water. Mass production of this parasitoid has been unsuccessful because it is a species with a high risk of death due to behavioral attributes and genetic degeneration in the colonization process (Nunes et al., 2012).

Fopius arisanus is an Asian parasitoid that lays eggs on first instar larvae of tephritids (Manoukis et al., 2011; Carmichael et al., 2005). In the late 1940s, it was introduced in Hawaii (USA) for control of *Bactrocera dorsalis* (Hendel) (Diptera: Tephritidae), where it also showed good development in *C. capitata* and thus can be used to control this species (Manoukis et al., 2011). After its success in Hawaii, *F. arisanus* was introduced for the control and suppression of fruit flies in Australia, Central America and several islands of the Pacific and Indian Oceans and in the Mediterranean basin (Vargas et al., 2001).

Estudos preliminares demostraram que de *F. arisanus* apresenta melhor desenvolvimento em ovos de *C. capitata* do que em ovos de *A. fraterculus*, indicando que *C. capitata* é o hospedeiro mais adequado para a criação massal do parasitoide visando a utilização em programas de controle aumentativo (Groth et al., 2016).

Another braconid used in biological control, *Psyttalia concolor* was introduced in the United States and Bolivia, however it was recovered only in very low numbers in the first country and never recovered in the second (Vaughan, 1992).

Diapriidae

The family Diapriidae, one of the largest families of Hymenoptera, belongs to the superfamily Proctotrupoidea and is divided into four subfamilies: Belytinae, Ismarinae, Ambositrinae and Diapriinae. The family contains about 150 genera and approximately 2,300 described species, the fauna being estimated at 4,500 species worldwide (Goulet; Huber, 1993).

The genus *Coptera* Say has been registered as parasitoid of several families of Diptera (Psilidae, Muscidae, Otitidae, Drosophilidae, Milichidae, Lonchaeidae), mainly Tephritidae. In the Neotropical region, the genus *Coptera* Say includes two species: *C. breviceps* (Kieffer) first recorded in the State of Pará (Brazil), and *C. haywardi* recorded in Tucumán (Argentina) attacking pupae of *Anastrepha fraterculus* and *A.*

schultzi (Loiácono 1981), in Mexico, attacking *A. obliqua* (Lopez et al., 1999), in Venezuela in *A. serpentina* and *A. striata* (Garcia; Montilla, 2001) and in Brazil (Aguiar-Menezes et al., 2003). Possibly this parasitoid should occur in other countries located between Argentina and Mexico, but it has not been registered for the lack of taxonomists. It is an idiobiont parasitoid of Tephritidae pupae and native to the Neotropical Region (Nuñez-Campero et al., 2012). Its high specificity provides it an advantage as a candidate for the biological control of flies of the genus *Anastrepha* spp. (Sivinski et al., 1988). In Argentina, *C. haywardi* is considered a valid option in the augmentative biological control of *A. fraterculus*, since its percentage of parasitism is high (about 7%) with a long period of oviposition (about 20 days) (Nuñez-Campero et al., 2012).

The parasitoid *C. haywardi* is more effective than *P. vindemiae* in pupae of fruit flies, while the second parasitize only pupae on the soil surface; *C. haywardi* parasitize pupae up to 5 cm deep. Thus, *C. haywardi* represents a viable candidate to replace *P. vindemiae* in augmentative biological control programs for fruit flies (Guillén et al., 2002).

The genus *Trichopria* includes microhymenopterans that can be used for the biological control of flies, since they are parasitoids of immature stages of Diptera. *Trichopria anastrephae* Lima is a generalist species, generally occurring a single parasitoid per puparium of the host (Ioriatti, 1995).

Lima (1940) described the species *T. anastrephae*, obtained from puparium of *Anastrepha serpentina* (Wiedemann) and *Anastrepha* sp. (Diptera: Tephritidae) originated in Rio de Janeiro. It is currently distributed in the states of Rio de Janeiro (Aguiar-Menezes et al., 2001), Minas Gerais (Silva et al., 2003), Goiás (Marchiori; Penteado-Dias, 2001), Santa Catarina (Garcia; Corseuil, 2004) and Rio Grande do Sul in *A. fra*terculus (Cruz et al., 2011). Moreover, it has already been registered in Argentina (Ovruski et al., 2004) and in Venezuela (Boscán; Godoy, 1996). This species is also a potential parasitoid of *Drosophila suzukii* (Wollmann et al., 2016).

Eulophidae

There are two species of Eulophidae that are parasitoids of *Anastrepha*: *Aceratoneuromyia indica* and *Tetrastichus giffardianus*. *Aceratoneuromyia indica* is a gregarious koinobiont parasitoid of *C. capitata* and *Anastrepha* spp. native to Southeast Asia. In turn, *Tetrastichus giffardianus* Silvestri, a parasitoid known of *C. capitata* for many years from West Africa, was recently recorded parasitizing *Anastrepha obliqua* (Macquart) (Diptera: Tephritidae) on fruits of umbu-cajazeira *Spondias* sp. (Anacardiaceae) in Brazil (Araújo et al., 2016).

Figitidae

Figitidae parasitoids constitute a group of very important parasitoids of Tephritidae, being one of the most diverse, and can be found in all biogeographical regions, although it is worth noting the diversity in tropical regions (Fontal-Cazalla et al., 2002).

The parasitoid *Aganaspis daci* is native to Malaysia and Borneo, was introduced in Hawaii on *B. dorsalis* (Hendel) (Clausen et al., 1965). In Florida (USA) for control of *Anastrepha suspensa* (Loew) (Baranowski et al., 1993), in Mexico (Jimenez-Jimenez, 1956) and Costa Rica (Wharton et al., 1981) for the biological control of *Anastrepha* spp.

Aganaspis pelleranoi can be mass-produced using *A. fraterculus* larvae for augmentative biological control programs. However, future studies should evaluate the effect of rearing for many generations under conditions of mass production (Gonçalves et al., 2016).

Pteromalidae

The family Pteromalidae is cosmopolitan containing more than 3,000 species distributed in 600 genera. It includes species of idiobiont and koinobiont parasitoids, ectoparasits, endoparasits, solitary and gregarious parasites and primary and secondary parasites (Hanson; Gauld, 1995).

Pachycrepoideus vindemmiae (Rondani) is a solitary parasitoid containing a wide host range in the Diptera families: Anthomyiidae, Calliphoridae, Muscidae, Sarcophagidae, Tachinidae and Tephritidae. This species is cosmopolitan and was found in North America and Africa (Marchiori et al., 2003). Other parasitoids of this family such as *Spalangia simplex*, *Spalangia cameroni* and *S. endius* are occasional parasitoids of *Anastrepha*.

In addition, it is considered a facultative hyperparasitoid, that is, capable of parasitizing many species of primary parasitoids such as *Fopius arisanus* (Sonan, 1932), *Diachasmimorpha longicaudata* (Ashmead, 1905) and *Psyttalia concolor* (Szépligeti, 1910) (Wang; Messing, 2004). Due to its low specificity and hyperparasitism, this parasitoid is not very suitable for integrating biological control programs of fruit flies.

Perspectives

The successful use of parasitoids for the suppression of fruit fly populations depends on studies that evaluate the behavior of these insects and the use of such methods in conjunction with other area-wide control techniques on an integrated pest management basis (IPM). Biological control and the sterile insect technique (SIT) are expected to be used across large areas, suppressing fly populations, reducing the source of infestations and the risk of spread of pest species, reducing the use of insecticides in fruit destined for domestic and foreign markets, and ultimately creating a biological barrier to fruit fly incursions in major horticultural regions (Garcia; Ricalde, 2013).

Studies seeking to integrate more than one type of biological control agent that exhibit synergistic effects are of fundamental importance within an attempt to increase the percentages of biological control efficiency.

The training of researchers and technicians, whether biologists or agronomists, with more expertise in Applied Ecology becomes fundamental, since beyond the knowledge of Entomology, it is fundamental to improve the conditions of agroecosystems, that is, agroecosystems should present better conditions to parasitoids, such as nectar, shelter, etc., thus allowing the maintenance of larger populations of biological control agents.

Niche modeling studies of natural enemies of fruit flies may greatly assist in defining where the best region for releasing a biological control agent is and which regions are potentially inadequate for the establishment of that organism.

Acknowledgments

We thank the Coordenação de Aperfeiçoamento de Pessoal de Nível Superior (Capes); Conselho Nacional de Desenvolvimento Científico e Tecnológico (CNPq) for the financial support and Instituto Nacional dos Hymenoptera Parasitoides (INCT-HYMPAR) for the financial support.

References

Aguiar-Menezes, E. L.; Menezes, E. B.; Silva, P. S.; Bittar, A. C. R.; Cassino, P.C.R. (2001). Native hymenopterous parasitoids associated with *Anastrepha* spp. (Diptera: Tephritidae) in Seropedica city, Rio de Janeiro, Brazil. *Flo. Entomol.* 84 (4): 706-711.

Aguiar-Menezes; E. L., Menezes, E. B.; Loiácono, M.S. (2003). First record of *Coptera haywardi* Loiácono (Hymenoptera: Diapriidae) as a parasitoid of fruit-infesting Tephritidae (Diptera) in Brazil. *Neotropical Entomol.* 32 (2): 355-358.

Aluja, M. Bionomics and management of *Anastrepha*. *Ann. Rev. Entomol.* 39: 155-173. 1994.

Aluja, M..; Sivinski, J..; Rull, J..; Hodgson, P. J. (2005). Behavior and predation of fruit fly larvae (*Anastrepha* spp.) (Diptera: Tephritidae) after exiting fruit in four types of habitats in tropical Veracruz, Mexico. *Environ. Entomol.* 34(6), 1507-1516.

Alves, S. B. Controle microbiano de insetos. Piracicaba: FEALQ, p. 1163, 1998. [Alves, S. B. Microbial control of insects. Piracicaba: FEALQ, p. 1163, 1998.]

Andaló, V; Nguyen, K; Moino Junior, A. (2006). *Heterorhabditis amazonensis* n. sp. (Rhabditida: Heterorhabditidae) from Amazonas, Brazil. *Nematol.* 8 (6): 853-867.

Araújo, A. A. R.; Silva, P. R. R.; Silva, R. B. Q.; Sousa, E. P. S. (2016). *Tetrastichus giffardianus* on pupae of *Anastrepha* in Brazil. *Ciênc. Rural*, 46 (7): 1134-1135.

Askew, R. R.; Shaw, M. R. Parasitoid communities: their size, structure and development In: Waage, J.; Greathead, D. *Insect parasitoids*. London: London Academic, 1986, p. 225-264.

Baker, A. C.; Stone, W. E.; Plummer, C. C.; Mcphail, M. (1944). A review of studies on the Mexican fruit fly and related Mexican species. *USDA M. Publ.*, 531, 1-155.

Ball, G. E.; Bousquet, Y. (2000) Carabidae latreille, 1810. In: Arnett, R. H., Thomas, M.C. (Eds.), *American beetles: Archostemata, Myxophaga, Adephaga, Polyphaga: Staphyliniformia*. (First Edition, 32-132) Boca Raton – Florida: CRC Press.

Baranowski, R.; Glenn, H.; Sivinski, J. (1993) Biological control of the Caribbean fruit fly (Diptera: Tephritidae). *Flo. Entomol.* 76: 245-251

Barbosa-Negrisoli, C. R. C; Garcia, M. S; Dolinski, C; Negrisoli Junior, A. S; Bernardi, D; Nava, D. E. (2009). Efficacy of indigenous entomopathogenic nematodes (Rhabditida: Heterorhabditidae, Steinernematidae), from Rio Grande do Sul, Brazil, against *Anastrepha fraterculus* (Wied.) (Diptera: Tephritidae) in peach orchards. *J. Invertebr. Pathol.* 102 (1): 6-13.

Batalla-Carrera, L.; Morton, A.; Garcia-del-Pino, F. (2016). Virulence of entomopathogenic nematodes and their symbiotic bacteria against the hazelnut weevil *Curculio nucum*. *J. Appl. Entomol.* 140 (1-2): 115–123.

Beavers, J. B.; Calkins, C. O. (1984). Susceptibility of *Anastrepha suspensa* (Diptera: Tephritidae) to Steinernematid and Heterorhabditid nematodes in laboratory studies. *Environ. Entomol.* 13 (1): 137-139.

Bel, Y. F.; Granero, T. M.; Alberola, M. J.; Martinez, S.; Ferre, J. (1997). Distribution, frequency and diversity of *Bacillus thuringiensis* in olive tree environments in Spain. *Appl. Microbiol.* 20 (1): 652-658.

Bortoli, L.; Machota, J. R. R.; Garcia, F. R. M.; Botton, M. (2016) Evaluation of food lures for fruit flies (Diptera: Tephritidae) captured in a Citrus orchard of the Serra Gaúcha. *Flo. Entomol.*, 99: 381-384.

Boscán, N. M; Godoy, F. (1996). Nuevos parasitoides de moscas de las frutas de los generos *Anastrepha* y *Ceratitis* en Venezuela. *Agronom. Trop.*, 4(4): 465-471. [Boscán, N.M; Godoy, F. (1996). New parasitoids of fruit flies of the genus *Anastrepha* and *Ceratitis* in Venezuela. *Agronom. Trop.*, 4 (4): 465-471.

Bravo, A.; Gillb, S. S.; Soberón, M. (2007). Mode of action of *Bacillus thuringiensis* Cry and Cyt toxins and their potential for insect control. *Toxicon*. 49 (4): 423-435.

Brida, A. L; Rosa, J. M. O; Oliveira, C. M. G; Castro, B. M. C; Serrão, J. E; Zanuncio, J. C; Leite, L. G; Wilcken, S. R. S. (2017). Entomopathogenic nematodes in agricultural areas in Brazil. *Sci. Rep.* 7 (45254): 1-7.

Brothers, D. J., Tschuch, G., Buger, F., (2000) Associations of mutillid wasp (Hymenoptera, Mutillidae) with eusocial insects. *Insect. Soc.*, 47, 201-211.

Buentello-Wong, S. L.; Galán-Wong, K.; V. Arévalo-Niño, V. Almaguer, C.; Rojas-Verde, G. (2015). Characterization of Cry Proteins in Native Strains of *Bacillus thuringiensis* and Activity against *Anastrepha ludens*. *Entomologist*. 40 (1): 15-24.

Campanini, E. B.; Davolos C. C.; Alves E. C.; Lemos, M. V. (2012). Isolation of *Bacillus thuringiensis* strains that contain Dipteran-specific cry genes from Ilha Bela (São Paulo, Brazil) soil samples. *Brazilian Journal of Biology*. 72 (1): 243-247.

Cancino, J.; Montoya, P. Advances and Perspectives in the mass rearing of fruit fly parasitoids in Mexico Fruit Flies of Economic Importance: From Basic to Applied Knowledge Proceedings of the 7th International Symposium on Fruit Flies of Economic Importance 10-15 September 2006, Salvador, Brazil pp. 133-142.

Canesin, A. (2011). Avaliação de nematoides entomopatogênicos (Rhabditida: Steinernematidae; Heterorhabditidae) no controle de moscas-das-frutas (Diptera: Tephritidae) e do gorgulho-da-goiaba (Coleoptera: Curculionidae). 2011. 116 f. Tese (Doutorado). Universidade Federal de Grande Dourados, Dourados, MG, 2011. URL: http://files.ufgd.edu.br/arquivos/arquivos/78/MESTRADO-DOUTORADO-AGRONOMIA/Tese%20Angela%20Canesin.pdf. [Canesin, A. (2011). Evaluation of entomopathogenic nematodes (Rhabditida: Steinernematidae; Heterorhabditidae) in control of fruit flies (Diptera: Tephritidae) and guava weevil (Coleoptera: Curculionidae). 2011. 116 f. Thesis (Doctorate). Federal University of Grande Dourados, Dourados, MG, 2011. URL: http://files.ufgd.edu.br/arquivos/arquivos/78/MESTRADO-DOUTORADO-AGRONOMIA/Tese%20Angela%20Canesin.pdf.]

Cantone, F. A.; Vandenberg, J. D. (1998). Intraspecific diversity in *Paecilomyces fumosoroseus*. *Mycological Research*. 102 (2): 209-215.

Carmichael, A. E.; Wharton, R. A.; Clarke, A. R. (2005) Opiine parasitoids (Hymenoptera: Braconidae) of tropical fruit flies (Diptera: Tephritidae) of the Australian and south pacific region. *Bull. Entomol. Res*. 95: 545–569.

Carneiro, G.; Salles, L. (1994). Patogenicidade de *Paecilomyces fumosoroseus*, isolado CG 260 sobre larvas e pupa de *Anastrepha fraterculus* Wied. *An. Soc. Entomol. Bras*. 23 (1): 341-343. [Carneiro, G.; Salles, L. (1994). Pathogenicity of *Paecilomyces fumosoroseus*, isolated CG 260 on larvae and pupa of *Anastrepha fraterculus* Wied. *An. Soc. Entomol. Bras*. 23 (1): 341-343]

Chesthukina, G. G.; Kostina, I. I.; Mikhailova, A. I.; Tyurin, S. A.; Klepikova, F. S.; Stepanov, V. M. (1982). The main features of *Bacillus thuringiensis* delta-endotoxin structure. *Arch. Microbiol.*132 (1): 159-162.

Clausen, C. P.; Clancy, D. W.; Chock, Q. C. Biological control of the oriental fruit fly (*Dacus dorsalis* Hendel) and other fruit flies in Hawai. Washington: United States Department of Agriculture Technical Bulletin, 1965.

Cole, L. (2003) Pesticide Programs, Prevention Division, Environmental Protection Agency, Washington, United States of American.

Costa, N. P.; Oliveira, H. D.; Brito, C. H.; Silva, A. B. (2007). Influência do Nim na biologia do predador *Euborellia annulipes* e estudos de parâmetros para sua criação massal. *R. Biol. C. Terra.* 7(2), 1-8. [Costa, N. P.; Oliveira, H. D.; Brito, C. H.; Silva, A.B. (2007). Influence of Nim on the biology of the predator *Euborellia annulipes* and studies of parameters for its mass rearing. *R. Biol. C. Terra.* 7(2), 1-8.]

Cruz, P. P.; Neutzling, A. S.; Garcia, F. R. M. (2011) Primeiro registro de *Trichopria anastrephae*, parasitoide de moscas-das-frutas, no Rio Grande do Sul. *Ciênc. Rur.*, 41: 1297-1299. [Cruz, P. P.; Neutzling, A. S.; Garcia, F. R. M. (2011) First record of *Trichopria anastrephae*, parasitoid of fruit flies, in Rio Grande do Sul. Ciênc. Rur., 41: 1297-1299.]

Destéfano, R. R. H.; Bechara, I. J.; Messias, C. L.; Piedrabuena, A. E. (2005). Effectiveness of *Metarhizium anisopliae* against immature stages of *Anastrepha fraterculus* fruit fly (Diptera: Tephritideae). *Braz. J. Microbiol.* 36(1): 94-99.

Dimbi, S.; Maniania, N. K.; Lux, S. A.; Mueke, J. M. (2003). Host Species, Age and Sex as Factors Affecting the Susceptibility of the African Tephritid Fruit Fly Species, *Ceratitis capitata*, *C. cosyra* and *C. fasciventris* to Infection by *Metarhizium anisopliae*. *J. Pest Sci.* 76 (1): 113-117.

Dimbi, S.; Maniania, N. K. (2009). Effect of *Metarhizium anisopliae* Inoculation on the Mating Behavior of Three Species of African Tephritid Fruit Flies, *Ceratitis capitata*, *Ceratitis cosyra* and *Ceratitis fasciventris*. *Biol. Control*. 50 (1): 111-116.

Dolinski, C; Kamitani, F. L; Machado, I. R; Winter, C. E. (2008). Molecular and morphological characterization of Heterorhabditid entomopathogenic nematodes from the tropical rainforest in Brazil. *Mem. Inst. Oswaldo Cruz.* 103 (2): 150–159.

Ekesi, S.; Maniania, N. K.; Mohamed, S. A.; Lux, S. A. (2005). Effect of soil application of different formulations of *Metarhizium anisopliae* on African Tephritide fruit flies and their associated endoparasitoids. *Biol. Control*, 35(1): 83-91.

Ekesi, S.; Maniania, N. K.; Lux, S. A. (2003). Mortality in three African tephritid fruit fly puparia and adults caused by the entomopathogenic fungi *Metarhizium anisopliae* and *Beauveria bassiana. Biocontrol Sci. Technol*, 12 (1): 7-17.

Eskafi, F. M.; Kolbe, M. M. (1990). Predation on larval and pupal *Ceratitis capitata* (Diptera: Tephritidae) by the ant *Solenopsis geminate* (Hymenoptera: Formicidae) and others predators in Guatemala. *Environ. Entomol.* 19 (1): 148-153.

Espósito, E.; Azevedo, J. L. Fungos: uma introdução à biologia, bioquímica e biotecnologia. Educs - Caxias do Sul, Coleção Biotecnologia, 510p. 2004. [Espósito, E.; Azevedo, J. L. Fungi: an introduction to biology, biochemistry and biotechnology. Educs - Caxias do Sul, Biotechnology Collection, 510p. 2004.]

Feng, M. C.; Poprawski, T. J.; Khachatourians, G. G. (1994). Production, formulation and application of the entomopathogenic fungus *Beauvaria bassiana* for insect control: current status. *Biocontrol Sci. Technol.* 4 (1): 3-34.

Fernandes, W. D.; Sant'Ana, M. V.; Raizer, J.; Lange, D. (2012). Predation of fruit fly larvae *Anastrepha* (Diptera: Tephritidae) by ants in grove. *Psyche*, 10: 83-89.

Ferraz, L. C. C. B. Nematoides entomopatogênicos. In: Alves, S.B (ed.). Controle Microbiano de Insetos. Fealq: Piracicaba, 1998, 541-569p. [Ferraz, L. C. C. B. Entomopathogenic nematodes. In: Alves, S.B (ed.). Microbial Control of Insects. Fealq: Piracicaba, 1998, 541-569p.

Ferry, N.; Edwards, M. G.; Gatehouse, J. A.; Gatehouse, A. M. R. (2004). Plant-insect interactions: molecular approaches to insect resistence. *Current Opinios in Biotecnology*. 15 (1): 155-161.

Ffrench-Constant, R. H. (2005) Something old, something transgenic, or something fungal for mosquito control?. *Ecol. Evol.* 20 (1): 577-579.

Foelkel, E.; Monteiro, B. L.; Voss, M. (2016). Virulence of nematodes against larvae of the south-American fruit fly in laboratory using soil from Porto Amazonas, Paraná, Brazil, as substrate. *Cienc. Rural.* 46 (3): 405–410.

Foelkel, E.; Voss, M.; Monteiro, L. B; Nishimura, G. (2017). Isolation of entomopathogenic nematode in na apple orchard in southern Brazil and its virulence to *Anastrepha fraterculus* (Diptera: Tephritidae) larvae, under laboratory conditions. *Braz. J. Biol.* 77 (1): 22-28.

Fontal-Cazalla, F. M.; Buffington, M. L.; Nordlander, G.; Liljeblad, J.; Ros-Farre, P.; Nieves-Aldrey, Y. J. L.; Pujade-Villar, J.; Ronquist, F.

(2002) Phylogeny of the Eucoilinae (Hymenoptera: Cynipoidea: Figitidae). *Cladistics*, 18 (2): 154 – 199.

Fuga, C. A. G; Fernandes, R. H; Lopes, E. A. (2012). Entomopathogenic nematodes. *Agr. Biol. Scien.* 6 (3): 56-75.

Fuxa, J. R. (1987). Ecological considerations for the use of entomopathogens in IPM. *Ann. Rev. Entomol.* 32 (1): 225-251.

Galli, J. C.; Rampazzo, E. F. (1996). Enemigos naturales predadores de *Anastrepha* (Diptera, Tephritidae) capturados con trampas de suelo em huertos de *Psidum guajava* L. *Bol. Sanid. Veg. Plagas*, 22: 297-300.

Garcia, F. R. M. Fruit fly: Biological and ecological aspects. In Current Trends In: Bandeira, R.R. *Fruit flies control on perennial crops and research prospects*; Kerala: Transworld Research Network, 2009; pp. 1–35.

Garcia, F. R. M.; Corseuil, E. (2004). Native hymenopteran parasitoids associated with fruit flies (Diptera: Tephritoidea) in Santa Catarina State, Brazil. *Flo. Entomol.*, 87 (4): 517-521.

Garcia, F. R. M.; Ricalde, M. P. (2013) Augmentative Biological Control using parasitoids for fruit fly management in Brazil. *Insects*, 4: 55-70.

García, J. L.; Montilla, R. (2001) *Coptera haywardi* Loiácono (Hymenoptera: Diapriidae) parasitoide de pupas de *Anastrepha* spp. (Diptera: Tephritidae) en Venezuela. *Entomotropica*, 16: 191-195 [García, J. L.; Montilla, R. (2001) *Coptera haywardi* Loiácono (Hymenoptera: Diapriidae) parasitoid of pupae of *Anastrepha* spp. (Diptera: Tephritidae) in Venezuela. *Entomotropica*, 16: 191-195.]

Glazer, I. Survival Biology. In: Gaugler, R. (ed.). Entomopathogenic Nematology, CAB International: New York, 2002, 169-187p.

Gonçalves, R. S.; Andreazza, F.; Lisboa, H.; Grutzmacher, A. D.; Valgas, R. A.; Manica-Berto, R.; Nornberg, S. D.; Nava, D. E. (2016). Basis for the development of a rearing technique of *Aganaspis pelleranoi* (Hymenoptera: Figitidae) in *Anastrepha fraterculus* (Tephritidae: Diptera). *J. Econ. Entomol.* 109: 1094-1101.

González, P. I.; Montoya, P.; Perez-Lachaud, G.; Cancino, J.; Liedo, P. (2007). Superparasitism in mass reared *Diachasmimorpha longicaudata* (Hymenoptera: Braconidae), a parasitoid of fruit flies (Diptera: Tephritidae). *Biol. Control*, 40 (3): 320-326.

Goulet, H.; Huber, J. T. Hymenoptera of the world: an identification guide to families. Toronto: Agriculture Canada Puclications, 1993.

Grewal, P. S; Nardo, E. A. B; Aguillera, M. (2001). Entomopathogenic nematode: Potential for exploration and use in South America. *Neotrop. Entomol.* 30 (2): 191-205.

Groth, M. Z.; Loeck, A. E.; Nornberg, S. D.; Bernardi, D.; Nava, D. E. (2016) Biology of *Fopius arisanus* (Hymenoptera: Braconidae) in two species of fruit flies. *J. Insect Sci.*, 16 (1): 96, 2016.

Guillén, L.; Aluja, M.; Equihua, M., Sivinski, J. (2002) Performance of two fruit fly (Diptera: Tephritidae) pupal parasitoids (*Coptera haywardi* [Hymenoptera: Diapriidae] and *Pachycrepoideus vindemiae* [Hymenoptera: Pteromalidae]) under different environmental soil conditions. *Biol. Control.* 23(3): 219-227.

Hajek, A. E.; Leger, R. J. (1994). Interactions between fungal pathogens and insect. *Ann Rev. Entomol.* 39 (1): 293-322.

Hall, R. A. (1982). Deuteromycetes: Virulence and bioassay desing. *Invertebr. Pathol.* 1 (1): 191-196.

Hanson, P. E.; Gauld, D. (1995). The Hymenoptera of Costa Rica. Oxford, Oxford University Press,

Hennessey, M. K. (1997). Predation on wandering larvae and pupae of Caribbean Fruit Fly (Diptera: Tephritidae) in Guava and Carambola Grove Soils. *J. Agr Entomol.*, 14(2), 129-138.

Hernández, M. A. R. (2003). Patogenicidad de nematodos entomopatógenos (Nematoda: Steinernematidae, Heterorhabdtidae) en larvas y pupas de mosca de la fruta *Anastrepha ludens* Loew (Diptera: Tephritidae) 112p. Universidad de Colima. Dissertação (Mestrado em Ciências), Universidad de Colima, México. URL: http://digeset.ucol.mx/tesis_posgrado/Pdf/Miguel%20Angel%20Reyes%20Hernandez.pdf. [Hernández, M.A. R. (2003). Pathogenicity of entomopathogenic nematodes (Nematoda: Steinernematidae, Heterorhabdtidae) in larvae and pupae of fruit fly *Anastrepha ludens* Loew (Diptera: Tephritidae) 112p. University of Colima. Dissertation (Master in Science.) URL: http://digeset.ucol.mx/tesis_posgrado/Pdf/Miguel%20Angel%20Reyes%20Hernandez.pdf.]

Heve, W. K; El-Borai, E. F; Carrillo, D; Ducan, L. W. (2016). Biological control potential of entomopathogenic nematodes for management of Caribbean fruit fly, *Anastrepha* suspensa Loew (Tephritidae). *Pest Manage. Sci.* 72.

Hofte, H.; Whiteley, R. H. (1989). Insecticidal crystal proteins of *Bacillus thuringiensis*. *Microbiol. Rev.* 53 (1): 242-255.

Holland, J. M.; Luff, M. L. (2000). The effects of agricultural practices on Carabidae in temperate agroecosystems. *Integrated. Pest. Manag. Rev.* 5(2): 109-129.

Ioriatti, M. C. S. S. Contribuição ao estudo da biologia e taxonomia de Hymenoptera parasitóides de Diptera das famílias Tephritidae e Lonchaeidade. 1995. 124f. Dissertação (Mestrado em Ecologia) - Universidade Federal de São Carlos, SP. [Ioriatti, M. C. S. S. Contribution to the study of the biology and taxonomy of Hymenoptera parasitoides of Diptera of the families Tephritidae and Lonchaeidade. 1995. 124f. Dissertation (Master in Ecology) - Federal University of São Carlos, SP.]

Jackson, M. A.; Mcguire, M. R.; Lacey, L. A. (1997) Liquid culture production of desiccation tolerant blastospores of the bioinsecticidal fungus *Paecilomyces fumosoroseus. Mycol.l Res.*. 101 (1): 35-41

.Jimenez-Jimenez, E. (1956) Las moscas de la fruta y sus enemigos naturales. *Fitófilo*, 16: 4-11. [Jimenez-Jimenez, E. (1956) Fruit flies and their natural enemies. *Fitófilo*, 16: 4-11.]

Kamali, S.; Karimi, J.; Hosseini, M.; Campos-Herrera, R.; Duncan, L. W. (2013). Biocontrol potential of the entomopathogenic nematodes *Heterorhabditis bacteriophora* and *Steinernema carpocapsae* on cucurbit fly, *Dacus ciliatus* (Diptera: Tephritidae). *Bio Sci. Technol.* 23 (11):1307–1323.

Kromp, B. (1999). Carabid beetles in sustainable agriculture: a review on pest control efficacy, cultivation impacts and enhancement. *Agric. Ecosyst. Environ.* 74 (1-3): 187-228.

Lewis, E. E.; Campbell, J.; Griyn, C.; Kaya, H.; Peters, A. (2006). Behavioral ecology of entomopathogenic nematodes. *Biol.Control.* 38 (1): 66-79.

Lezama-Gutiérrez, R.; Trujillo de la Luz, A.; Molina-Ochoa, J.; Rebolledo-Domínguez, O.; Pescador, A. R.; Lopez-Edwars, M.; Aluja, M. (2000). Virulence of *Metarhizium anisopliae* (Deuteromycotina: Hyphomycetes) on *Anastrepha ludens* (Diptera: Tephritidae) laboratory and field trials. *J. Economic Entomol.* 93 (1): 1080-1084.

Lezama-Gutiérrez, R.; Molina-Ochoa, J.; Pescador-Rubio, A.; Galindo-Velasco, E.; Ángel-Sahagún, C. A.; Michel-Aceves, A. C.; González-

Reyes, E. (2006). Efficacy of Steinernematid nematodes (Rhabditida: Steinernematidae) on the suppression of *Anastrepha ludens* (Diptera: Tephritidae) larvae in soil of differing textures: laboratory and field trials. *J Agri. Urban Entomol.* 23 (1): 41-49.

Lezama-Gutiérrez, R; Molina, J.; Contreras, Oscar, L. (1996) Susceptibilidad de larvas de *Anastrepha ludens* (Diptera: Tephritidae) a diversos nemátodos entomopatógenos (Steinernematidae y Heterorhabditidae). *Vedalia* 3: 31-34. [Lezama-Gutiérrez, R; Molina, J; Contreras, Oscar, L. (1996) Susceptibility of larvae of *Anastrepha ludens* (Diptera: Tephritidae) to various entomopathogenic nematodes (Steinernematidae and Heterorhabditidae). *Vedalia* 3: 31-34.]

Liedo, P.; Leon, E.; Barrios, M. I.; Valle-Mora, J. F.; Ibarra, G. (2002). Effect of age on the mating propensity of the Mediterranean fruit fly (Diptera: Tephritidae). *Fla Entomol.* 85 (1): 94-101.

Lima, A. C. (1940) Alguns parasitos de moscas de frutas. *An. Acad. Bras. Ciênc.* 12 (1): 17-20. [Lima, A.C. (1940) Some parasites of fruit flies. *An. Acad. Bras. Ciênc.* 12 (1): 17-20.]

Lima, R. L.; Andreazze, R.; Andrade, H. T. A.; Pinheiro, M. P. G., (2010). Riqueza de famílias e hábitos alimentares em Coleoptera capturados na Fazenda da EMPARN – Jiqui, Parnamirim/RN. *Entomobrasilis*. 3(1), 11-15. [Lima, R. L.; Andreazze, R.; Andrade, H.T.A.; Pinheiro, M.P., G., (2010). Richness of families and dietary habits in Coleoptera captured at EMPARN - Jiqui Farm, Parnamirim / RN. *Entomobrasilis*. 3 (1), 11-15.]

Loiácono, M. S. (1981) Notas sobre *Diapriinae neotropicales* (Hymenoptera, Diapriidae). *Rev. Soc. Entomol. Argentina*. 40: 237-241. [Loiácono, M. S. (1981) Notes on neotropical Diapriidae (Hymenoptera, Diapriidae). *Rev. Soc. Entomol. Argentina*. 40: 237-241.]

López, M., M. Aluja, AND J.; Sivinski. (1999) Hymenopterous larval-pupal and pupal parasitoids of *Anastrepha* flies (Diptera: Tephritidae) in Mexico. *Biol. Control* 15: 119-129.

Lovei, G. L.; Sunderland, K. D. (1996). Ecology and behavior of ground beetles (Coleoptera: Carabidae). *Annu. Rev. Entomol.*, 41 : 231-256.

Malavasi, A.; Zucchi, R. A.; Sugayama, R. L. Biogeografia. In: Malavasi, A.; Zucchi, R.A. *Moscas das-frutas de importância econômica no Brasil.* Ribeirão Preto: Holos 2000, p.93-98.

[Malavasi, A.; Zucchi, R.A.; Sugayama, R. L. Biogeography. In: Malavasi, A.; Zucchi, R.A. *Fruit flies of economic importance in Brazil*. Ribeirão Preto: Holos 2000, p.93-98.]

Mangan, R. L.; Moreno, D. S.; Thompson, G. D. (2006) Bait dilution, spinosad concentration, and efficacy of GF-120 based fruit fly sprays. *Crop Protection*. 25: 125- 133.

Manoukis, N.; Geib, S.; Seo, D.; Mckenney, M.; Vargas, R.; Jang, E. (2011). An optimized protocol for rearing *Fopius arisanus*, a parasitoid of tephritid fruit flies. *J. Vis. Exp.* 53: 1–4.

Marchiori, C. H.; Pereira, L. A.; Silva Filho, O. M. (2003) Primeiro relato do parasitóide *Pachycrepoideus vindemmiae* (Rondani) (Hymenoptera: Pteromalidae) parasitando pupas de *Sarcodexia lambens* Wiedemann (Diptera: Sarcophagidae). *Cienc. Rural*, 33:173-175. [Marchiori, C.H.; Pereira, L. A.; Silva Filho, O. M. (2003) First report of the parasitoid *Pachycrepoideus vindemmiae* (Rondani) (Hymenoptera: Pteromalidae) parasitizing pupae of *Sarcodexia lambens* Wiedemann (Diptera: Sarcophagidae). *Cienc. Rural*, 33: 173-175.]

Marchiori, C. H; Penteado-Dias, A. M. (2001) *Trichopria anastrephae* (Hymenoptera: Diapriidae), parasitóide de Diptera, coletadas em área de mata nativa e pastagem em Itumbiara, Goiás, Brasil. *Arq. Inst. Biol.* 68 (1): 123-124. [Marchiori, C. H.; Penteado-Dias, A. M. (2001) *Trichopria anastrephae* (Hymenoptera: Diapriidae), parasitoid of Diptera, collected in native forest area and pasture in Itumbiara, Goiás, *Brazil. Inst. Biol.* 68 (1): 123-124.]

Martinez, A. J.; Robacker, D. C.; Garcia, J. A. (1997). Toxicity of an isolate of *Bacillus thuringiensis* subspecies *darmstadiensis* to adults of the Mexican fruit fly (Diptera: Tephritidae) in the laboratory. *J. Econ. Entomol.* 90 (1): 130- 134.

McClintock, J. T.; Schaffer, C. R.; Sjoblad, R. D. (1995). A comparative review of the mammalian toxicity of *Bacillus thuringiensis*-based pesticides. *Pestic. Sci*. 45 (1): 95-105.

Meirelles, R. N.; Redaelli, L. R.; Ourique, C. B. (2013) Comparative biology of *Diachasmimorpha longicaudata* (Hymenoptera: Braconidae) reared on *Anastrepha fraterculus* and *Ceratitis capitata* (Diptera: Tephritidae). *Flo. Entomol*. 96: 412-418.

Meirelles, R. N.; Redaelli, L. R.; Ourique, C. B; Jahnke, S. M. (2016) Parasitismo de *Anastrepha fraterculus* por *Diachasmimorpha longicaudata* em condições de semicampo. *Agrária*. 11: 204-209. [Meirelles, R. N.; Redaelli, L. R.; Ourique, C.B; Jahnke, S. M. (2016) Parasitism of *Anastrepha fraterculus* by *Diachasmimorpha longicaudata* in semi-field conditions. *Agrária*. 11: 204-209.]

Meyling, N. V.; Eilenberg, J. (2007) Ecology of the entomopathogenic fungi *Beauveria bassiana* and *Metarhizium anisopliae* in temperate agroecosystems: Potential for conservation biological control. *Biol. Cont.* 43 (1): 145-155.

Morgante, J. S. Moscas-das-frutas (Tephritidae): características biológicas, detecção e controle, Brasília: SENIR, 1991 [Morgante, J. S. Fruit flies (Tephritidae): biological characteristics, detection and control, Brasília: SENIR, 1991.]

Nguyen, K. B; Ginarte, C. M. A; Leite, L. G; Santos, J. M; Harakava, R. (2010). *Steinernema brazilense* n. sp. (Rhabditida: Steinernematidae), a new entomopathogenic nematode from Mato Grosso, Brazil. *J. Invertebr. Pathol.* 3 (1): 8–20.

Nguyen, K. B; Shapiro-Ilan, D. I; Fuxa, J. R; Wood, B. W; Bertolotti, M. A; Adams, B. J. (2006). Taxonomic and biological characterization of *Steinernema rarum* found in the Southeastern United States. *J. Nematol.* 38 (1): 28–40.

Norrbom. Tephritidae classification table. 2012/11/10. Avaiable from: http://www.sel.barc.usda.gov/diptera/tephriti/tephclas.htm.

Novelo-Rincón, L. F.; Montoya, P.; Hernández-Ortiz, V.; Liedo, P.; Toledo, J. (2009). Mating performance of sterile Mexican fruit fly *Anastrepha ludens* (Dipt., Tephritidae) males used as vectors as *Beauveria bassiana* (Bals.) Vuill. J. *App. Entomol.* 133 (1): 702–710.

Nunes, A. M.; Nava, D. E.; Müller, F. A.; Gonçalves, R. S.; Garcia, M. S. (2012) Biology and parasitic potential of *Doryctobracon areolatus* on *Anastrepha fraterculus* larvae. *Pesq. Agropec. Bras.* 46: 669–671.

Nuñez-Campero, S. R.; Ovruski, S., Aluja, M. (2012) Survival analysis demographic parameters ofthe pupal parasitoid *Coptera haywardi* (Hymenoptera, Diapriidae), reared on *Anastrepha fraterculus* (Diptera: Tephritidae). *Biol. Control*, 61: 40-46.

Ohba, M.; Aizawa, K. (1986). Distribution of *Bacillus thuringiensis* in soils of Japan. *J. Invertebr. Pathol.* 47 (1): 277-282.

Ovruski, S. M.; Aluja, M.; Sivinski, J.; Wharton, R. A. (2000) Hymenopteran parasitoids on fruit-infesting Tephritidae (Diptera) in

Latin America and the Southern United States: diversity, distribution, taxonomic status and their use in fruit fly biological control. *Int. Pest Manag. Rev*. 5(2): 81-107.

Ovruski, S. M.; Schlisermana, P.; Aluja, M. (2004) Indigenous parasitoids (Hymenoptera) attacking *Anastrepha fraterculus* and *Ceratitis capitata* (Diptera: Tephritidae) in native and exotic host plants in Northwestern Argentina. *Biol. Control*. 29 (1): 43-57.

Perez, M.; Toledo, J.; Liedo, P. (2000). Fecundity and Ovarian Development in Females of *Anastrepha obliqua* (Macquart) with Four Food Sources. *Flo. Entomol*. 108 (1): 43-51.

Poinar Jr, G. O. (1975). Entomogenous nematodes, a manual and host list of insect-nematode associations. Leiden: E.J. Brill, 254p.

Poinar, G. O.; Grewal, P. S. (2012). History of entomopathogenic nematology. *J. Nematol*. 44(2): 153–161.

Polanczyk, R. A.; Alves, S. (2003) *Bacillus thuringiensis*: Uma breve revisão. *Agrociência*. 7(1): 1-10. [Polanczyk, R.A.; Alves, S. (2003) *Bacillus thuringiensis*: A brief review. *Agrociencia*. 7 (1): 1-10.]

Polanczyk, R. A.; Valicente, F. H.; Barreto, M. R. (2008). Utilização de *Bacillus thuringiensis* no controle de pragas agrícolas na América Latina, p.111-136. In: Alves, S.B.; Lopes, R.B. (Eds.). *Controle microbiano de pragas na América Latina: avanços e desafios*. Piracicaba, FEALQ, p. 414. [Polanczyk, R.A.; Valicente, F. H.; Barreto, M. R. (2008). Use of *Bacillus thuringiensis* in agricultural pest control in Latin America, p.111-136. In: Alves, S.B.; Lopes, R.B. (Eds.). *Microbial pest control in Latin America: advances and challenges*. Piracicaba, FEALQ, p. 414.]

Quesada-Moraga, E.; Ruiz-Garcia, A.; Santiago-Alvarez, C. (2006). Laboratory evaluation of entomopathogenic fungi *Beauveria bassiana* and *Metarhizium anisopliae* against puparia and adults of *Ceratitis capitata* (Diptera: Tephritidae). *J. Economic Entomol*. 99 (1): 1955-1966.

Quesada-Moraga, E.; Carballo-Martin, I.; Jurado-Garrido, I.; Alvarez-Santiago, C. (2008). Horizontal transmission of *Metarhizium anisopliae* among laboratory populations of *Ceratitis capitata* (Wiedemann) (Diptera: Tephritidae). *Biol. Control*. 47 (1): 115-124.

Radeghieri, P. (2004). *Cameraria ohridella* (Lepidoptera Gracillaridae) predation by *Crematogaster scutellaris* (Hymenoptera Formicidae) in Northern Italy (Preliminary note). *B. Insectol.* 57(1): 63-64.

Rafael, J. A.; Melo, G. A. R.; de Carvalho, C. J. B.; Constantino, R. (2012). *Insetos do Brasil, Diversidade e Taxonomia*. Ribeirão Preto: Holos. [Rafael, J.A.; Melo, G.A. R.; De Carvalho, C.J.B.; Constantino, R. (2012). *Insects of Brazil, Diversity and Taxonomy*. Ribeirão Preto: Holos.]

Raga, A.; Sato, M. E. (2005) Effect of spinosad bait against *Ceratitis capitata* (Wied.) and *Anastrepha fraterculus* (Wied.) (Diptera: Tephritidae) in laboratory. *Neotrop. Entomol.*, 34: 815-822.

Ricklefs, R. E. A economia da natureza. 6 Edition. Rio de Janeiro: Editora Guanabara Koogan; 2010. [Ricklefs, R. E. The economy of nature. 6 Edition. Rio de Janeiro: Guanabara Koogan Publishing House; 2010.]

Robacker, D. C.; Martinez, A. J.; Garcia, J. A.; Diaz, M.; Romero, C. (1996). Toxicity of *Bacillus thuringiensis* to Mexican fruit fly (Diptera: Tephritidae). *J. Econ. Entomol.* 89 (1): 104-110.

Rodrigues-Trentine, F. (1996). Mecanismos de defesa e controle de *Anastrepha fraterculus* (Wiedemann, 1830) (Diptera: Tephritidae) expostas a nematoides entomopatogênicos. Curitiba: Universidade Federal do Paraná, 73p. Dissertação de Mestrado em Ciências Biológicas. [Rodrigues-Trentine, F. (1996). *Anastrepha fraterculus* (Wiedemann, 1830) defense and control mechanisms (Diptera: Tephritidae) exposed to entomopathogenic nematodes. Curitiba: Federal University of Paraná, 73p. Master's Dissertation in Biological Sciences.]

Rosa, W.; Lopez, F.; Liedo, P. (2002). *Beauveria bassiana* as a pathogen of the mexican fruit fly (Diptera: Tephritidae) *J. Economic Entomol.* 95(1): 36-43.

Sánchez-Roblero, D.; Huerta-Palacios, G.; Valle, J.; Gómez, J.; Toledo, J. (2012). Effect of *Beauveria bassiana* on the ovarian development and reproductive potential of *Anastrepha ludens* (Diptera: Tephritidae). *Biocontrol Sci. Technol.* 22 (9): 1075-1091.

Schnepf, E.; Crickmore, N.; Rie, J.V.; Lereclus, D.; Baum, J.; Feitelson, J.; Zeigler, D. R.; Dean, D. H. (1998). *Bacillus thuringiensis* and its pesticidal crystal proteins. *Microbiology and Molecular Biology Review*. 62 (1): 775-806.

Scoz, P. L.; Botton, M.; Garcia, M. S. (2004) Controle químico de *Anastrepha fraterculus* (Wied.) (Diptera: Tephritidae) em laboratório. *Ciênc, Rural.* 30: 1689-1684. [Scoz, P. L.; Botton, M.; Garcia, M. S. (2004) Chemical control of *Anastrepha fraterculus* (Wied.) (Diptera: Tephritidae) in laboratory. *Ciênc, Rural.* 30: 1689-1684]

Silva, A. B.; Batista, J. L.; Brito, C. H. (2010), Capacidade predatória de *Euborellia annulipes* (Dermaptera: Anisolabididae) sobre *Hyadaphis foeniculi* (Hemiptera: Aphididae). *R. Bras. Biol. C. Terra.* 10, 44-51. [Silva, A. B.; Batista, J.L.; Brito, C. H. (2010), Predatory capacity of *Euborellia annulipes* (Dermaptera: Anisolabididae) on *Hyadaphis foeniculi* (Hemiptera: Aphididae). *R. Bras. Biol. C. Terra.* 10, 44-51.]

Silva, C. G.; Marchiori, C. R.; Fonseca, A. R., Torres, L. C. (2003) Himenópteros parasitóides de larvas de *Anastrepha* spp. em frutos de carambola (*Averrhoa carambola* L.) na região de Divinópolis, Minas Gerais, Brasil. *Ciênc. Agrotec.* 27(6): 1264-1267. [Silva, C. G.; Marchiori, C.R.; Fonseca, A. R., Torres, L. C. (2003) Hymenopteran parasitoids of larvae of *Anastrepha* spp. In fruits of carambola (*Averrhoa carambola* L.) in the region of Divinópolis, Minas Gerais, Brazil. *Ciênc. Agrotec.* 27 (6): 1264-1267]

Sivinski, J.; Aluja, M.; López, M. (1997) Spatial and temporal distributions of parasitoids of Mexican *Anastrepha* species (Diptera: Tephritidae) within the canopies of fruit trees. *Ann. Entomol. Soc. Am.* 90: 604-618.

Sivinski, J.; Vulinec, K. E.; Aluja, M. (1998) The bionomics of *Coptera haywardi* (Oglobin) (Hymenoptera: Diapriidae) and other pupal parasitoids of tephritid fruit flies (Diptera). *Biol. Control.* 11: 193-202.

Sivinski, J. M.; Calkins, C. O.; Baranowski, R.; Harris, D.; Brambila, J.; Diaz, J.; Burns, R. E.; Holler, T.; Dodson, G. (1996). Supression of a Caribbean Fruit Fly (*Anastrepha suspense*) population through augmented releases of the parasitoid *Diaschasmimorpha longicaudata. Biol. Control.* 6(2), 177-185.

Starnes, R. L.; Liu, C. L.; Marrone, P. G. (1993). History, use and future of microbial insecticides. *Am. Entomol.* 39 (1): 83-91.

Stibick, J. N. L. (2004). Natural enemies of true fruit flies (Tephritidae). United States Department of Agriculture Animal and Plant Health Inspection service Plant Protection and Quarantine.

Thomas, D. B. (1995). Predation on the soil inhabiting stages of the Mexican fruit fly. *Southwest. Entomol.*,20, 60-71.

Thompson, J. N. The Coevolutionary Process. Chicago, Illinois, United States of America: University of Chicago Press; 1994.

Toledo, J.; Pérez, C.; Liedo, P.; Ibarra, J. E. (2005a). Susceptibilidad de larvas de *Anastrepha obliqua* Macquart (Diptera: Tephritidae) a *Heterorhabditis bacteriophora* (Poinar) (Rhabditida: Heterorhabditidae) en condiciones de laboratorio. *Vedalia*. 12 (1): 11-22 [Toledo, J.; Pérez, C.; Liedo, P.; Ibarra, J. E. (2005a). Susceptibility of larvae of *Anastrepha obliqua* Macquart (Diptera: Tephritidae) to *Heterorhabditis bacteriophora* (Poinar) (Rhabditida: Heterorhabditidae) under laboratory conditions. *Vedalia*. 12 (1): 11-22.]

Toledo, J.; Campos, S. E.; Flores, S.; Liedo, P.; Barrera, J. F.; Villasenor, A.; Montoya, P. (2007). Horizontal Transmission of *Beauveria bassiana* in *Anastrepha ludens* (Diptera: Tephritidae) under laboratory and field cage Conditions. *J. Economic Entomol*. 100 (1): 292-297.

Toledo, J.; Flores, S.; Montoya, P. (2010). Insectos esteriles como vectores de patogenos in moscas de la fruta: Fundamentos y procedimientos para su manejo, eds. P. Montoya p: 369- 376. [Toledo, J.; Flores, S.; Montoya, P. (2010). Sterile insects as vectors of pathogens in fruit flies: Fundamentals and procedures for their management, eds. P. Montoya p: 369-376.]

Toledo, J.; Liedo, P.; Williams, T.; Ibarra, J. (1999). Toxicity of *Bacillus thuringiensis* beta-exotoxin to three species of fruit flies (Diptera: Tephritidae). *J. Econ. Entomol.* 92 (1): 1052-1056.

Toledo, J.; Sánchez, J. E; Willians, T; Gómez, A; Montoya, P; Ibarra, J. E. (2014). Effect of soil moisture on the persistence and efficacy of *Heterorhabditis bacteriophora* (Rhabditida: Heterorhabditidae) against *Anastrepha ludens* (Diptera: Tephritidae) larvae. *Flo. Entomol.* 97 (2): 528-533.

Toledo, J; Gurgúa, J. L; Liedo, F. P; Ibarra, J. E; Oropeza, A. C. (2001). Parasitismo de larvas y pupas de la mosca mexicana de la fruta, *Anastrepha ludens* (Loew) (Diptera: Tephritidae) por el nemátodo *Steinernema feltiae* (Filipjev) (Rhabditida: Steinernematidae). *Vedalia*. 8: 27-36. [Toledo, J; Gurgúa, J. L; Liedo, F. P; Ibarra, J.E.; Oropeza, A.C. (2001). Parasitism of larvae and pupae of the Mexican fruit fly, *Anastrepha ludens* (Loew) (Diptera: Tephritidae) by the nematode

Steinernema feltiae (Filipjev) (Rhabditida: Steinernematidae). *Vedalia.* 8: 27-36.]

Toledo, J; Ibarra, J. E; Liedo, P; Gomez, A; Rasgado, M. A; Williams, T. (2005a). Infection of *Anastrepha ludens* (Diptera: Tephritidae) larvae by *Heterorhabditis bacteriophora* (Rhabditida: Heterorhabditidae) under laboratory and field conditions. *Bio. Sci. Technol.* 15 (6): 627-634.

Toledo, J; Rasgado, M. A; Ibarra, J. E; Gomez, A; Liedo, P; Williams, T. (2006a) Infection of *Anastrepha ludens* following soil applications of *Heterorhabditis bacteriophora* in a mango orchard. *Entomol. Exp. Appl.* 119 (2): 155-162.

Toledo, J; Rojas, R; Ibarra, J. E. (2006b). Efficiency of *Heterorhabditis bacteriophora* (Nematoda: Heterorhabditidae) on *Anastrepha serpentina* (Diptera: Tephritidae) larvae under laboratory conditions. *Flo. Entomol.* 89 (4): 524-526.

Townsend, C. R.; Begon, M.; Harper, J. L. (2006). Predação, pastejo e doenças. In: Townsend, C. R., Begon, M., & Harper, J. L. (Eds.), *Fundamentos em Ecologia*, (2 Edition, 293-331). Porto Alegre: Artmed. [Townsend, C.R.; Begon, M.; Harper, J.L. (2006). Predation, grazing and diseases. In: Townsend, C.R., Begon, M., & Harper, J.L. (Eds.), *Fundamentals in Ecology*, (2 Edition, 293-331). Porto Alegre: Artmed.]

Tumialis, D. Pezowicz, E.; Skrzecz, I.; Mazurkiewicz, A.; Maszewska, J.; Pietraszczylk, J. J; Kuchasrska, K. (2016). Occurrence of entomopathogenic nematodes in Polish soils. *Cienc. Rural.* 46 (7): 1126–1129.

Vallete-Gelly, I.; Lemaitre, B.; Boccard, F. (2008). Bacterial strategies to overcome insect defenses. Nature Reviews: *Microbiol.* 6 (4): 302-313.

Vargas, R. I.; Ramadan, M.; Hussain, T.; Mochizuki, T.; Bautista, R. C.; Stark, J. D. (2001) Comparative demography of six fruit fly (Diptera: Tephritidae) parasitoids (Hymenoptera: Braconidae). *Biol. Control.* 25: 30–40.

Vaughan, M. A. International biocontrol cooperation within Latin America. In Coulson; J.R.; M.C. Zapater, M.C. *Opportunities for implementation of biocontrol in Latin America*, Buenos Aires: SRNT-IOBC, 1992, p. 7–38.

Vilas-Bôas, G. T.; Peruca, A. P. S.; Arantes, O. M. N. (2007). Biology and taxonomy of *Bacillus cereus*, *Bacillus anthracis* and *Bacillus thuringiensis*. *Can. J. Microbiol.* 53 (1): 673-687.

Wang X. G.; Messing, R. H. (2004). The ectoparasitic pupal parasitoid, *Pachycrepoideus vindemmiae* (Hymenoptera: Pteromalidae); attacks other primary tephritid fruit fly parasitoids: host expansion and potential non-target impact. *Biol. Control*, 31: 227-236.

Wharton, R. A.; Gilstrap, F. E.; Rhodei, R. H.; Fischel, M. M.; Hart, W. G. (1981) Hymenopterous egg-pupal and larval-pupal parasitoids of *Ceratitis capitata* and *Anastrepha* spp. (Diptera: Tephritidae) in Costa Rica. *Entomophaga*, 26: 285-290.

Wharton, R. A.; Marsh, P. M.; Sharkey, M. J. Manual of the New World Genera of the Family Braconidae (Hymenoptera). London: International Society of Hymenopterists, 1997.

Wollmann, J.; Schlesener, D. C. H.; Ferreira, M. S.; Garcia, M. S.; Costa, V. A.; Garcia, F. R. M. (2016) Parasitoids of Drosophilidae with potential for parasitism on *Drosophila suzukii* in Brazil. *Dros. Infor. Serv.* 99: 38-42, 2016.

Yamazaki, K., Sugiura, S. (2006). Feeding of a shore-inhabiting ground beetle, *Scarites aterrimus* (Coleoptera: carabidae). *Coleopts. Bull.* 60(1), 75-79.

Zucchi, R. A. Diversidad, distribución y hospederos del género *Anastrepha* en Brasil, In: Hernández-Ortiz, V. *Moscas de la fruta em Latinoamérica (Diptera: Tephritidae): diversidad, biologia y manejo.* Distrito Federal: S y G editores, 2007, p.77-100. [Zucchi, R. A. Diversity, distribution and hosts of the genus *Anastrepha* in Brazil, In: Hernández-Ortiz, V. *Fruit flies in Latin America (Diptera: Tephritidae): diversity, biology and management*. Federal District: S y G editors, 2007, p.77-100.]

In: Biological Control
Editor: Lewis Davenport

ISBN: 978-1-53612-416-3

Chapter 3

Biological Control and Pesticides: Integration in Greenhouse Production Systems

Raymond A. Cloyd*
Department of Entomology, Kansas State University,
Manhattan, KS, US

Abstract

Biological control is a plant protection strategy widely-used in horticultural cropping systems to regulate insect and mite pest populations on greenhouse-grown ornamentals and vegetables. However, in certain instances, biological control may not provide sufficient regulation of insect and mite pest populations. Therefore, pesticides (insecticides and miticides) are oftentimes required to provide supplemental suppression of pest populations. The integration of natural enemies or biological control agents (parasitoids and predators) and pesticides is a strategy that is gaining interest among greenhouse

* Corresponding author address. Email: rcloyd@ksu.edu.

producers due to the potential for reduced pesticide inputs, which may result in decreased development of pesticide resistance in insect and mite pest populations, and potential economic and environmental benefits. This chapter discusses the advantages and issues affiliated with using natural enemies in conjunction with pesticides and provides insights on the practicality of using both plant protection strategies simultaneously. Furthermore, the direct and indirect effects of entomopathogenic fungi, botanical pesticides, insecticidal soaps and horticultural oils, and fungicides on natural enemies are discussed. Finally, guidelines for enhancing the integration of biological control and pesticides are presented.

Keywords: biological control agents, natural enemies, entomopathogenic fungi, botanical pesticides, insecticidal soaps, horticultural oils, fungicides

INTRODUCTION

Biological control is a preventative plant protection strategy used in greenhouse production systems associated with greenhouse-grown horticultural crops, including ornamentals and vegetables. Natural enemies or biological control agents, such as parasitoids and predators, are released augmentatively to regulate insect or mite pest populations (van Lenteren 2012, Cloyd 2012a). However, natural enemies will not "eradicate" an insect or mite pest population. Therefore, the success of natural enemies is contingent on regulating or maintaining insect or mite pest numbers at levels low enough to prevent or minimize plant damage (Cloyd 2012a, Gigon et al. 2016).

INTEGRATING NATURAL ENEMIES WITH PESTICIDES

Natural Enemies and Pesticides

First, we need to discuss the natural enemies and pesticides used in greenhouse production systems. There are two natural enemy insect or mite types that may be purchased and released into greenhouses: parasitoids and predators (Albajes et al. 1999, Heinz et al. 2004). In general, parasitoids used in greenhouses attack (parasitize) either one or several insect hosts and/or a particular host life stage (egg, larva, nymph, pupa, or adult). Parasitoids are free-living in the adult stage while the immature life stage (larva) is either inside (endo-parasitoid) or outside (ecto-parasitoid) the host (Osborne et al. 2004, Cloyd 2012a). An individual female parasitoid inserts an egg(s) into a host, the egg(s) hatches into a larva that consumes the internal contents of the host, the larva transitions into a pupa, and then eventually emerges as an adult parasitoid (Osborne et al. 2004). Predators are commonly generalists, feeding on more than one host type and life stage of a host. In most instances, the larva or nymph and adult are predaceous feeding on a wide-variety of different hosts (Osborne et al. 2004, Cloyd 2012a).

The pesticides used in greenhouse production systems that may directly or indirectly affect natural enemies are insecticides, miticides, and fungicides (Cloyd 2012b). Pesticides (insecticides and miticides) are classified as contact, stomach poison, translaminar, or systemic (Ware and Whitacre 2003, Cloyd 2011). Furthermore, pesticides may be either broad-spectrum or narrow-spectrum in their activity. Broad-spectrum pesticides are active on many different insect and mite pests, and may be harmful to natural enemies. However, narrow-spectrum pesticides, also called selective pesticides, are active on specific insect or mite pests and are generally less harmful to natural enemies (Ripper et al. 1949, Croft 1990, Cloyd 2012b). The direct and indirect effects of selective pesticides, including: insect growth regulators and selective feeding blockers on natural enemies have been thoroughly discussed by Cloyd 2012b.

Therefore, this chapter focuses on the challenges and opportunities associated with integrating pesticides and natural enemies in relation to the direct and indirect effects of entomopathogenic fungi, botanical pesticides, insecticidal soaps and horticultural oils, and fungicides on natural enemies.

Integration of Natural Enemies with Pesticides

Greenhouse producers are implementing biological control programs in greenhouses primarily due to issues related to insect or mite pest populations developing pesticide resistance, increased costs and regulations affiliated with registering new active ingredients, and potential negative effects on pollinators (Shipp et al. 1991, Osborne and Oetting 1989, van Lenteren 2000, Wawrzynski et al. 2001, Iwasa et al. 2004, Chauzat et al. 2006, Pilkington et al. 2010). However, there are cases where natural enemies fail to establish themselves; are not commercially available; quality is inadequate; or their use is too expensive (Osborne and Oetting 1989, Parrella et al. 1992, Stevens et al. 2000, Heinz et al. 2004, Fiedler and Sosnowska 2007, Messelink et al. 2014, Gonzalez et al. 2016). In addition, the use of natural enemies alone may not be sufficient to effectively regulate insect or mite pest populations within a greenhouse, and/or reduce subsequent plant damage (Osborne and Oetting 1989, Stark et al. 2007). Therefore, pesticides may be needed to provide supplemental suppression of existing pest populations in order to prevent damage to greenhouse-grown horticultural crops (van den Bosch and Stern 1962, Hayashi 1996, Stark and Banks 2003, Hassan and Van de Veire 2004, Stark et al. 2007).

There is a general interest among greenhouse producers in integrating, combining, or simultaneously using natural enemies and pesticides (insecticides and miticides) so as to successfully suppress insect or mite pest populations (Croft and Brown 1975, Parrella 1999, Nicetic et al. 2001). The potential benefits of integrating natural enemies and pesticides include: reduced pesticide inputs (based on frequency of applications), which may lower the risk of pesticide resistance developing in insect and mite pest populations; provide a safer work environment for employees; and minimize pesticide residues on plant material (Gentz et al. 2010). Moreover, the integration of natural enemies and pesticides may provide effective long-term suppression of pest populations compared to either strategy used alone (Sosnowska and Piatkowski 1996, Lacey et al. 2001). However, the effects of pesticide applications on the life cycle and efficacy of natural enemies are complex (Holt and Hochberg 1997, Gentz et al. 2010), and there may be instances were pesticides and natural enemies cannot be used together.

The current terminology that describes the possibility of utilizing natural enemies and pesticides together is somewhat confusing. The term "compatibility", for example, has been used in numerous scientific and non-scientific publications (Parrella 1999, Jacobson et al. 2001, Williams et al. 2003, Hassan and Van de Veire 2004, Cloyd et al. 2006, Stark et al. 2007, Medina et al. 2007, Labbé et al. 2009, Cloyd 2012b, Bernardi et al. 2013, Saito and Brownbridge 2016, Mar Fernández et al. 2017); however, the term does not define what impact pesticides may have on natural enemies (e.g., survival) from a quantitative perspective. For instance, what level of mortality (based on percent) of natural enemies, when exposed to pesticides constitutes "compatibility?" Even natural enemy mortality levels <50% can substantially delay parasitoid population growth for one-generation, and consequently impact regulation of pest populations (Stark et al. 2007). Moreover, natural enemy mortality levels >50% are a concern in regards to the establishment of natural enemies and subsequent regulation of pest populations (Stark et al. 2007). Furthermore, there are issues affiliated with differences in methodology when assessing "compatibility" (Parrella 1999). Otieno et al. (2017) indicated that "compatibility" was associated with 'causing mortality (<20%)' although there was no indication what this level of natural enemy mortality refers to.

Therefore, when assessing the overall impact of pesticides on natural enemies, the terms direct and indirect may be more appropriate, thus reducing the extraneous terminology, and allowing for a more quantitative assessment in determining the possibility of integrating natural enemies with pesticides (Kim et al. 2006, Cloyd 2012b). In reality, greenhouse producers want to integrate pesticides with natural enemies in order to regulate insect or mite pest populations over the long-term without directly or indirectly affecting any life history parameters (e.g., predation, parasitism, longevity or survival, reproduction, and foraging behavior) or population dynamics (interplay between populations and environmental factors) of natural enemies (Stark and Banks 2003, Desneux et al. 2007, Cloyd 2012a), which will optimize regulation of pest populations. Consequently, both direct and indirect effects may influence the success of integrating natural enemies with pesticides by negatively affecting life history parameters and/or population dynamics (Desneux et al. 2007, Gonzalez et al. 2016).

Direct and Indirect Effects of Pesticides on Natural Enemies

Direct effects are associated with acute mortality or survival (longevity), over a specified time period, of the life stages of natural enemies including: egg, larva, nymph, pupa, or adult (Yardim and Edwards 1998, Cloyd 2012b, Amarasekare and Shearer 2013). Indirect effects are those that negatively impact the physiology or behavior of natural enemies by means of inhibiting reproduction, development time, sex ratio, mobility, searching and feeding behavior, and/or levels of parasitism or predation (Moriarty 1969, Haynes 1988, Theiling and Croft 1988, Desneux et al. 2007, Gentz et al. 2010). Any indirect effects on these factors alone or combinations may inhibit the ability of natural enemies to effectively regulate pest populations (Gonzalez et al. 2016). In addition, the use of pesticides may result in a depletion of the pest population, which serves as a food source for natural enemies; consequently, the pest population may be too low to sustain a natural enemy population (Yardim and Edwards 1998, Elzen 2001, Van Driesche and Heinz 2004, Devine and Furlong 2007, Cloyd and Bethke 2011), thus leading to natural enemies migrating away from treated areas (Newsom 1967).

Indirect effects of pesticides on natural enemies may be affiliated with differences in exposure based on feeding behavior. For example, predators such as ladybird beetles with chewing mouthparts that consume the entire host may be exposed to pesticide residues that persist on the pest exoskeleton (skin). However, predators, including predatory bugs (e.g., *Orius* spp.) that have sucking mouthparts may be less exposed due to their consuming internal host tissues (Theiling and Croft 1988).

Parasitoids, in general, are more sensitive to certain pesticides than predators (Theiling and Croft 1988, Sugiyama et al. 2011), which may be associated with indirect effects including: increased development time and/or decreased reproduction (Williams et al. 2003, Gentz et al. 2010). Nonetheless, any direct and/or indirect effects of pesticides on parasitoids and predators will vary depending on stage of development (egg, larva, pupa, or adult) exposed to pesticide residues (Stevenson and Walters 1983, Elzen 1990, Gentz et al. 2010, Amarasekare and Shearer 2013). What follows in the proceeding sections are examples of how entomopathogenic fungi, botanical pesticides, insecticidal soaps and horticultural oils, and fungicides may or may not directly or indirectly affect natural enemies.

Entomopathogenic Fungi

Entomopathogenic fungi are relatively ubiquitous world-wide (Starnes et al. 1993, Kanzok and Jacobs-Lorena 2006) and are commonly used in greenhouse production systems to suppress populations of different insect pests, such as: aphids, thrips, and whiteflies (Hall and Burges 1979, Shipp et al. 2003, Fiedler and Sosnowska 2007, Kivett et al. 2015). Entomopathogenic fungi, in general, infect the insect cuticle by means of enzymatic degradation and/or mechanical pressure (Burges and Hussey 1971, Cherwonogrodzky 1980, Gillespie and Claydon 1989, Clarkson and Charnley 1996, Gonzalez et al. 2016). Once inside the host, the entomopathogenic fungus may be distributed throughout the haemocoel, which is the primary body cavity that contains circulatory fluids (Clarkson and Charnley 1996). Death normally occurs in 3 to 14 days after the conidia (spores) of the entomopathogenic fungus contact the host (Gillespie and Claydon 1989). Host mortality may be caused by mechanical damage via spore penetration, resulting in water loss, and/or poisonous toxins produced by the entomopathogenic fungus (Cherwonogrodzky 1980, Gillespie and Claydon 1989). Mortality is generally dose-dependent, with higher conidia concentrations leading to faster kill and enhanced mortality rates (Vestergaard et al. 1995, James et al. 1998).

There are a number of pesticides containing entomopathogenic fungi as the active ingredient that are commercially available for use in greenhouse production systems, including: *Beauveria bassiana* (Balsamo-Crivelli) Vuillemin (BotaniGard®: BioWorks, Inc.; Victor, NY), *Metarhizum anisopliae* (Metschnikoff) Sorokin (Met52®: Monsanto BioAg, Inc.; St. Louis, MO), and *Isaria fumosorosea* (Wize) Brown & Smith (Preferal™: SePRO Corp.; Carmel, IN). The direct and indirect effects of entomopathogenic fungi on natural enemies is contingent on the following factors: 1) product formulation, 2) spore concentration, 3) natural enemy type (parasitoid or predator) and species, 4) life stage exposed (egg, larva, nymph, pupa, or adult), 5) timing of application (spatially and temporally), and 6) environmental conditions (temperature, relative humidity, and light intensity) (Lacey et al. 1977, Stevenson and Walters 1983, Mar Fernández et al. 2017). Furthermore, biological parameters that may be indirectly affected by exposure to entomopathogenic fungi include: host acceptance, reproduction, foraging behavior, sex ratio, and host emergence (for parasitoids) (Cloyd 2012b).

The integration of natural enemies and entomopathogenic fungi may be influenced by food availability and avoidance factors. For instance, any changes in host population numbers due to applications of an entomopathogenic fungus may reduce availability of food sources (Gonzalez et al. 2016), thus indirectly affecting subsequent natural enemy populations (Rosenheim et al. 1995). In addition, predators may avoid consuming hosts that are infected by entomopathogenic fungi (Meyling and Pell 2006, Labbé et al. 2009, Pourian et al. 2011, Seiedy et al. 2013). Entomopathogenic fungi produce volatiles after infecting hosts, which may subsequently mask the volatiles emitted by hosts that are used by predators to determine their location (Seiedy et al. 2013). In all three instances prey search time increases, predation rates decrease, and efficacy associated with regulating pest populations is reduced (Meyling and Pell 2006, Labbé et al. 2009, Pourian et al. 2011).

Additional factors that need to be considered in regards to the direct and indirect effects of entomopathogenic fungi on natural enemies are: 1) host treated with an entomopathogenic fungus may not be acceptable as a food source for parasitoids or predators (Mesquita and Lacey 2001); 2) parasitoids may avoid laying eggs into hosts already infected by an entomopathogenic fungus (Brobyn et al. 1988, Fransen and van Lenteren 1993); 3) entomopathogenic fungi may be able to outcompete immature parasitoids developing inside hosts (Furlong and Pell 2000, Furlong and Pell 2005); 4) parasitoids may become infected by an entomopathogenic fungus while developing inside an infected host (Furlong and Pell 2005); 5) during searching, parasitoids and predators may encounter free conidia of entomopathogenic fungi on plants and subsequently become infected (Roy and Pell 2000, Furlong and Pell 2005); 6) predators may ingest spores of an entomopathogenic fungus when consuming an infected host, thus compromising efficacy of both the predator and entomopathogenic fungus (Rosenheim et al. 1995, Longley and Stark 1996, Pell et al. 1997, Roy and Pell 2000); 7) a parasitoid larva developing inside a host may inadvertently ingest fungal spores (Askary and Brodeur 1999); and 8) multiple-pest complexes may influence the interactions between natural enemies and entomopathogenic fungi (Parrella 1990, Wawrzynski et al. 2001, Sugiyama et al. 2011, Gigon et al. 2016).

Natural enemies may ingest fungal spores when grooming or when feeding on contaminated hosts, although any indirect effects will depend on the concentration of viable spores present (Broadbent and Pree 1984, James and Lighthart 1994, Cloyd 2006). In addition, predators may not accept a host infected with an entomopathogenic fungus or the predator may ingest spores of an entomopathogenic fungus while consuming a host (Furlong and Pell 2005, Labbé et al. 2006, Cloyd 2012b). Furthermore, direct or indirect effects of entomopathogenic fungi on parasitoids may be contingent on interference competition where an entomopathogenic fungus infects a host before a parasitoid (Furlong and Pell 2005). The exclusion of the parasitoid, based on the entomopathogenic fungus having a temporal (time) advantage, may compromise host acceptability (Brooks 1993). An entomopathogenic fungus may also out-compete a parasitoid for host resources resulting in host death before the parasitoid can complete development (Powell 1986). However, if a parasitoid has a temporal (time) advantage, then the parasitoid can develop successfully (Furlong and Pell 2005). Therefore, temporal separation of natural enemies and entomopathogenic fungi may lead to successful suppression of insect pest populations (King and Bell 1978).

Smith and Krischik (2000) demonstrated a delayed effect associated with adult survival of the ladybird beetle, *Cryptolaemus montrouzieri* Mulsant (Coleoptera: Coccinellidae), when exposed to *B. bassiana* with 100% survival after 8 hours of exposure but 10% survival after 72 hours of exposure. Another study found that 50% of *C. montrouzieri* larvae died after consuming mealybugs contaminated with *B. bassiana* (Kiselek 1975). However, a comprehensive study demonstrated that *B. bassiana* was not directly harmful (based on percent infection) to adults of the predatory mites, *Neoseiulus* (formally = *Amblyseius*) *cucumeris* Oudemans (Arachnida: Phytoseiidae), and *Phytoseiulus persimilis* Athias-Henriot (Acarina: Phytoseiidae), and larvae of the parasitoids, *Eretmocerus eremicus* Rose & Zolnerowich (Hymenoptera: Aphelinidae), and *Aphidius colemani* Viereck (Hymenoptera: Aphidiidae) (Shipp et al. 2003). In addition, there were no direct effects to adults of the parasitoids, *Encarsia formosa* Gahan (Hymenoptera: Aphelinidae), and *E. eremicus*, and the predatory midge, *Aphidoletes aphidimyza* (Rondani) (Diptera: Cecidomyiidae), when exposed to *B. bassiana*; whereas, *A. colemani* adults were directly affected by *B. bassiana* (Shipp et al. 2003). A 7 day exposure period was not directly harmful to nymphs or adults of *N. cucumeris* (Jacobson et al. 2001). Labbé et al. (2009) reported that exposure to *B. bassiana* did not indirectly affect the parasitism rate of *E. formosa*, on the greenhouse whitefly, *Trialeurodes vaporariorum* (Westwood) (Hemiptera: Aleyrodidae). However, a 72 hour exposure period to *B. bassiana* reduced adult survival and reproduction of the predatory mite, *Neoseiulus californicus* (McGregor) (Acari: Phytoseiidae), with 71.2% mortality of adult females (Castagnoli et al. 2005a).

Shipp et al. (2012) reported that *B. bassiana* was harmful to adults of the predatory bug, *Orius insidiosus* (Say) (Hemiptera: Anthocoridae), whereas Ludwig and Oetting (2001) found that exposure to *B. bassiana*, under laboratory conditions, did not directly affect *O. insidiosus*, adults. Herrick and Cloyd (2017) indicated that *B. bassiana* and *M. anisopliae* were not directly harmful to adult *O. insidiosus*, with 100% and 80% survival, respectively. Any discrepancies associated with the direct effects of entomopathogenic fungi on natural enemies may be due to differences in application rate (concentration) or exposure period (Herrick and Cloyd 2017).

Botanical Pesticides

Botanical pesticides are those derived from plants, which includes extracts and oils (Isman 2000, Koul et al. 2008, Isman and Grieneisen 2014). The use of botanical pesticides may enhance the effectiveness of natural enemies due to the minimal residual activity associated with botanical pesticides (Parrella 1999, Koul et al. 2008). However, botanical pesticides are not widely-used in greenhouse production systems. Nevertheless, there are several botanical pesticides derived from the neem tree, *Azadirachta indica* A. Jusse, including: azadirachtin and clarified hydrophobic extract of neem oil (Saxena 1989, Otieno et al. 2017) that are used in greenhouse production systems. Azadirachtin is an insect growth regulator, acting as an ecdysone antagonist by delaying the molting process of larvae (Sieber and Rembold 1983, Schmutterer 1990, Mordue et al. 1998, Ware and Whitacre 2003). Since insect growth regulators are primarily active on the immature (e.g., larvae) life stages of insects (Vestergaard et al. 1995, Tunaz and Uygun 2004), any direct or indirect effects on the adult stage of natural enemies should be minimal (Biondi et al. 2012). Castagnoli et al. (2005b) reported that, under laboratory conditions, exposure to azadirachtin had no direct or indirect effects (based on survival and reproduction) on adults of the predatory mite, *N. californicus*. However, application rates of 3% and 5% of azadirachtin were highly toxic to the predatory mite, *P. persimilis*, with >77% mortality of nymphs and adults after 72 hours of exposure (Papaioannou-Souliotis et al. 2000).

The predatory mites, *P. persimilis* and *N. cucumeris*, were not directly or indirectly affected, based on percent adult mortality and eggs laid per female, after 24 and 48 hours of exposure to azadirachtin, respectively (Spollen and Isman 1996). However, larvae and adult females of the aphid predator, *A. aphidimyza,* were directly affected when exposed to azadirachtin with 35% mortality of the 1st instar larvae after 4 hours of exposure (Spollen and Isman 1996). Castagnoli et al. (2002) found that the predatory mite, *Amblyseius andersoni* Chant (Acari: Phytoseiidae), was not directly harmed when exposed to azadirachtin with >90% adult survival after 72 hours. Azadiracthin was also not directly harmful (based on percent survival) to *N. cucumeris* adults after 24 and 48 hours of exposure (Oetting and Latimer 1995).

Fecundity or egg hatch of the predatory green lacewing, *Chrysoperla carnea* Stephens (Neuroptera: Chrysopidae), were not negatively affected when exposed to azadirachtin (Medina et al. 2004). However, azadirachtin, at high concentrations (96 mg active ingredient/L), was directly harmful (98.2% mortality) to *C. carnea* adults and there were indirect effects on female oviposition (Medina et al. 2003). Smith and Krischik (2000) found that azadiracthin was not directly harmful to the predatory ladybird beetle, *C. montrouzieri*, after 72 hours of exposure. In addition, azadirachtin, under laboratory conditions, did not affect predation of the predatory rove beetle, *Dalotia* (formally = *Atheta*) *coriaria* (Kraatz) (Coleoptera: Staphylinidae), adults on fungus gnat, *Bradysia* sp. nr. *coprophila* (Linter) (Diptera: Sciaridae), larvae after 48 hours of exposure (Cloyd et al. 2009). In general, 24 hours of exposure to azadirachtin pure and formulated product, under laboratory conditions, was not directly harmful to the natural enemies, *A. colemani*, *A. aphidimyza*, and *N. cucumeris* (Stara et al. 2011). Rothwangl et al. (2004) reported that a 48 hour exposure period to azadirachtin did not directly affect the ability of the parasitoid, *Leptomastix dactylopii* (Howard) (Hymenoptera: Encyrtidae), to parasitize the citrus mealybug, *Planococcus citri* (Risso) (Hemiptera: Pseudococcidae).

Angeli et al. (2005) reported that a 7 day exposure period to azadirachtin did not affect the survival or fecundity of the predatory bug, *Orius laevigatus* Fieber (Hemiptera: Anthocoridae). Exposure to 1 hour, 7 day, and 14 day old residues of azadirachtin resulted in <40% mortality of *O. laevigatus* (Biondi et al. 2012). Herrick and Cloyd (2017) found that azadirachtin (Molt-X®: BioWorks, Inc.; Victor, NY) was not directly harmful to *O. insidiosus* with 100% adult survival up to 96 hours of exposure.

Capsicum oleoresin extract, garlic oil, and soybean oil (Captiva®: Gowan Comp.; Yuma, AZ) was not directly harmful to *O. insidiosus*, with 80% adult survival after 96 hours of exposure (Herrick and Cloyd 2017). A *Chenopodium*-based botanical pesticide (currently sold as Requim®: Bayer CropScience; Research Triangle Park, NC) was not directly harmful to *O. insidiosus*, adults and nymphs, under laboratory conditions, with 38.3% and 10.9% mortality, respectively, after 48 hours of exposure (Bostanian et al. 2005). In addition, the same botanical pesticide was not directly harmful to *A. colemani*, with 14.2% mortality after 48 hours of exposure (Bostanian et al. 2005).

Insecticidal Soaps and Horticultural Oils

Insecticidal soaps are derived from potassium salts of fatty acids (Cloyd 2004, Yu 2008) and studies have been conducted regarding their direct and indirect effects on natural enemies. Insecticidal soap was found to be highly toxic to the predatory bug, *O. insidiosus* adults and adults of the aphid parasitoid, *A. colemani*, with percent mortality >99% after 24 hours of exposure for *O. insidiosus*, and >86% for *A. colemani*, after 5 days of exposure (Bostanian and Akalach 2004). Bostanian and Akalach (2006) found that the predatory mites, *P. persimilis* and *Amblyseius fallacis* (Garman) (Acari: Phytoseiidae), and *O. insidiosus* were directly affected by insecticidal soap after 7 days of exposure with 99.7%, 83.2%, and

99.5% mortality, respectively. In another study, Bostanian et al. (2005) demonstrated that insecticidal soap was directly harmful to *O. insidiosus* adults and nymphs with >99% mortality after 48 hours of exposure. Moreover, in the same study, insecticidal soap was directly harmful to *A. colemani* adults with 63.4% mortality after 48 hours of exposure (Bostanian et al. 2005). However, Puritch et al. (1982) reported that exposure to insecticidal soap was not directly harmful to the parasitoid, *E. formosa* with 81.5% adult survival and there were no direct effects associated with residues on treated plants.

Oetting and Latimer (1995) found that an insecticidal soap formulation was directly harmful to the predatory mite, *N. cucumeris* after 24 hours of exposure with 90% mortality. Slide-dip tests, using an insecticidal soap, resulted in 100% mortality of *P. persimilis* adults; however, the ability of the predatory mite to feed on twospotted spider mite, *Tetranychus urticae* Koch (Acari: Tetranychidae), adults was not indirectly affected when plants (*Brassaia actinophylla* Endl.) were sprayed with the insecticidal soap 3 days after releasing the predatory mites (Osborne and Petitt 1985). A study found that insecticidal soap (potassium salts of fatty acids: M-Pede®; Mycogen, San Diego, CA) was not directly harmful to adults (based on percent survival) of the predatory ladybird beetle, *C. montrouzieri*, after 72 hours of exposure (Smith and Krischik 2000).

Horticultural oils used in greenhouse production systems are either petroleum, paraffinic, mineral, or neem-based (Cloyd 2016). Horticultural oils have been reported to kill natural enemies on contact when exposed to directed sprays (Davidson et al. 1991). Two horticultural oils were found to be directly harmful to the predatory mite, *N. cucumeris* adults with >80% mortality after 24 hours of exposure (Oetting and Latimer 1995). However, Nicetic et al. (2001) found that two applications of horticultural oil (petroleum-based), at 7 day intervals, did not directly affect populations of *P. persimilis* in the lower and upper canopies of greenhouse-grown roses (*Rosa* spp.). Moreover, Smith and Krischik (2000) demonstrated that 72 hours of exposure to a horticultural oil (paraffinic oil: Sunspray Ultrafine Oil®; Sun, Philadelphia, PA) was not directly harmful (based on percent survival) to adults of the predatory ladybird beetle, *C. montrouzieri*. Biondi et al. (2012) reported that a mineral oil did not directly harm *O. laevigatus* adults with <30% mortality after 3 days of exposure to 1 hour, 7 day, and 14 day old residues.

Fungicides

Fungicides are widely used in greenhouse production systems to control fungal diseases, and natural enemies may be inadvertently exposed to spray applications (Cloyd 2012b). Some fungicides may be integrated with natural enemies (van Lenteren 2000). In addition, databases (e.g., SELECTV) have indicated that fungicides are not directly harmful to natural enemies (Wright and Verkerk 1995) and there is a general assumption, overall, that fungicides can be integrated with natural enemies (Theiling and Croft 1988). Nonetheless, fungicides may have either direct or indirect effects on natural enemies (Tauber and Helgesen 1978, Yardim and Edwards 1998, Jepsen et al. 2007).

Studies have found that the older (developed prior to 1980), broad-spectrum fungicides may negatively affect certain natural enemies (Livingston et al. 1978, Roger et al. 1994, Bower et al. 1995). Fiedler and Sosnowska (2014) reported that the broad-spectrum fungicide, chlorothalonil was directly harmful to *P. persimilis* adults, under laboratory conditions, with ≥77% mortality 5 days after exposure. However, Malezieux et al. (1992) demonstrated, under both laboratory and field conditions that ten older, broad spectrum fungicides were not directly harmful to the larva and adult of *P. persimilis*. The broad spectrum fungicide, thiophanate-methyl was not directly harmful to *A. swirskii* with 18% adult mortality after a 1 day exposure period (Fiedler and Sosnowska 2014). In addition, Biondi et al. (2012) reported that two sulfur-based fungicides were not directly harmful (based on percent mortality) to *O. laevigatus* adults. Sugiyama et al. (2011) also found that a sulfur-based fungicide was not directly harmful (based on percent mortality) to the parasitoids; *Eretmocerus mundus* Mercet (Hymenoptera: Aphelinidae), *E. eremicus*, and *E. formosa* after 24 hours of exposure.

The newer (developed after 1980), more site-specific fungicides may be less harmful to natural enemies than older fungicides. For instance, Gradish et al. (2011) reported that newer fungicides, such as myclobutanil (Nova® 40W: Dow Agro-Sciences Canada; Calgary, Alberta), potassium bicarbonate (MilStop®: BioWorks USA; Victor, NY), and cyprodil + fludioxonil (Switch® WG: Syngenta Crop Protection Canada; Guelph, Ontario) were not directly harmful to *O. insidiosus* adults. Hogendorp and Cloyd (2013) found that, under laboratory conditions, potassium bicarbonate was not directly harmful to *L. dactylopii* and *C. montrouzieri* adults after 96 hours of exposure (percent mortality ≤30%) to rates ≤3.5 g/L. However, rates ≥5.5 g/L were more directly harmful, after 96 hours of exposure, to *C. montrouzieri* than *L. dactylopii* (90% vs. 0% mortality at 12.0 g/L). In a study by Herrick and Cloyd (2017), none of the currently available fungicides tested [aluminum tris (Aliette®: Bayer CropScience; Research Triangle Park, NC), azoxystrobin (Heritage®: Syngenta Crop Protection, Inc.; Greensboro, NC), fenhexamid (Decree®: SePRO Corp.; Carmel, IN), and kresoxim-methyl (Cygnus®: BASF Corp.; Research Triangle Park, NC)] were directly harmful to *O. insidiosus* with 100% to 80% adult survival after 96 hours of exposure, and there were no indirect effects on predation of western flower thrips, *Frankliniella occidentalis* (Pergande) (Thysanoptera: Thripidae), adults.

How to Integrate Biological Control and Pesticides

The success of integrating natural enemies with pesticides (insecticides, miticides, and fungicides) depends on spatial (space) and temporal (time) separation, which are associated with the timing of natural enemy releases and timing of pesticide applications (King and Bell 1978, van Lenteren and Woets 1988, Hassan and Van de Veire 2004, Biondi et al. 2012, Hogendorp and Cloyd 2013, Fiedler and Sosnowska 2014, Gonzalez et al. 2016). For instance, Cloyd et al. (2009) found that delaying releases of rove beetle, *D. coriaria*, adults after 48 hours following application of insecticides to the growing medium resulted in a substantial increase (approximately 60%) in adult survival compared to releasing 24 hours after applying the insecticides. The application of pesticides when the "most-tolerant" life stage (e.g., egg or pupa) is present may reduce any direct or indirect effects on natural enemies (Stevenson and Walters 1983, Ruberson et al. 1998, Hassan and Van de Veire 2004, Biondi et al. 2015). However, this would require extensive scouting efforts in order to ascertain the presence of the "most-tolerant" life stages prior to application of pesticides (Hassan and Van de Viere 2004, Roubos et al. 2014) and oftentimes more than one life stage is present at any given time (Cloyd 2016). In addition, the use of selective insecticides may reduce any direct or indirect effects on natural enemies (Bartlett 1964, Elzen et al. 1989, Schneider et al. 2004, Desneux et al. 2007, Sugiyama et al. 2011, Roubos et al. 2014, Biondi et al. 2015).

It is important to note that the exposure period (time) and rate (concentration) tested, under both laboratory and greenhouse conditions, may have a significant influence on the direct and indirect effects of pesticides on natural enemies as higher rates or concentrations tend to be more directly or indirectly harmful (Oetting and Latimer 1995, Bostanian et al. 2005, Hogendorp and Cloyd 2013, Fiedler and Sosnowska 2014, Mar Fernández et al. 2017). Furthermore, caution must be exercised when attempting to translate laboratory bioassays into predictions associated with the effects of pesticides on the performance of natural enemies in greenhouses. For instance, laboratory bioassays may fail to take into account the indirect effects of pesticides, which could underestimate their overall impact (Wright and Verkerk 1995, Amarasekare and Shearer 2013). Therefore, long-term evaluations conducted under greenhouse conditions provide more practical and applicable information regarding pesticide-pest-natural enemy interactions (Stark et al. 1995, Villanuéva-Jimenez and Hoy 1998). Moreover, the effects of pesticides on searching behavior and repellent effects of pesticides are difficult to quantify in laboratory bioassays (Parrella 1999). However, pesticides that have no direct or indirect effects on natural enemies in laboratory bioassays are unlikely to have any effects under greenhouse conditions (Wright and Verkerk 1995, Stark et al. 2007).

So, what are the possibilities of integrating natural enemies with pesticides and is the plant protection strategy feasible? Greenhouse producers need to address three key questions prior to integrating pesticides with natural enemies; these questions are the following:

1) Is a pesticide needed or will the natural enemies regulate pest populations enough to avoid noticeable plant damage?
2) Will the pesticide application enhance or disrupt the normal regulatory process provided by natural enemies?
3) Is the integration of natural enemies with pesticides practical and cost effective?

Conclusion

The integration of biological control agents or natural enemies with pesticides (insecticides and miticides) may be a viable plant protection strategy to deal with insect and mite pests of greenhouse production systems. However, the strategy is complex due to the various interactions between pesticides and natural enemies (Gentz et al. 2010). Furthermore, there are issues that have been presented in this chapter, which must be considered prior to integrating both plant protection strategies. For instance, any delay in the population growth of natural enemies by a pesticide can impact the ability of natural enemies to successfully regulate pest populations (Stark et al. 2007). In addition, certain pesticides (even fungicides) can disrupt natural enemies by means of direct or indirect effects (Tauber and Helgeson 1978), thus resulting in increased problems with insect or mite pests. Therefore, spatial separation associated with the timing of natural enemy releases and pesticide applications, and the use of selective insecticides, will help to successfully integrate both plant protection strategies.

ACKNOWLEDGMENTS

The author wants to thank Dr. Nathan J. Herrick of the Department of Entomology at Kansas State University (Manhattan, KS) and Dr. Mary Beth Kirkham of the Department of Agronomy at Kansas State University for reviewing initial drafts of the book chapter.

REFERENCES

Albajes, R., Gullino, M.L., van Lenteren, J.C., Elad, Y. editors. Integrated pest and disease management in greenhouse crops. The Netherlands: Kluwer: Dordrecht; 1999.

Amarasekare, K.G., Shearer, P., 2013. Laboratory bioassays to estimate the lethal and sublethal effects of various insecticides and fungicides on *Deraeocoris brevis* (Hemiptera: Miridae). *J. Econ. Entomol*. 106, 776-785.

Angeli, G., Baldessari, M., Maines, R., C. Duso, C., 2005. Side-effects of pesticides on the predatory bug *Orius laevigatus* (Heteroptera: Anthocoridae) in the laboratory. *Biocontrol Sci. Techn*. 15, 745-754.

Askary, H., Brodeur. J., 1999. Susceptibility of larval stages of the aphid parasitoid *Aphidius nigripes* to the entomopathogenic fungus, *Verticillium lecanii*. *J. Invertebr. Pathol*. 73, 129-132.

Bartlett, B.R. Integration of chemical and biological control. In DeBach P, editor. Biological Control of Insect Pests and Weeds. New York, NY: Reinhold; 1964; 489-514

Bernardi, D., Botton, M., Silva da Cunha, U., Bernardi, O., Malausa, T., Silveira Garcia, M., Nava. D.E., 2013. Effects of azadirachtin on *Tetranychus urticae* (Acari: Tetranychidae) and its compatibility with predatory mites (Acari: Phytoseiidae) on strawberry. *Pest Manag. Sci*. 69, 75-80.

Biondi, A., Campolo, O., Desneux, N., Siscaro, G., Palmeri, V., Zappalà, L., 2015. Life stage-dependent susceptibility of *Aphytis melinus*

DeBach (Hymenoptera: Aphelinidae) to two pesticides commonly used in citrus orchards. *Chemosphere* 128, 142-147.

Biondi, A., Desneux, N., Siscaro, G., and Zappalá, L., 2012. Using organic-certified rather than synthetic pesticides may not be safer for biological control agents: selectivity and side effects of 14 pesticides on the predator *Orius laevigatus*. *Chemosphere* 87, 803-812.

Bostanian, N.J., Akalach, M., Chiasson., H., 2005. Effects of a *Chenopodium*-based botanical insecticide/acaricide on *Orius insidiosus* (Hemiptera: Anthocoridae) and *Aphidius colemani* (Hymenoptera: Braconidae). *Pest Manag. Sci.* 61, 979-984.

Bostanian, N.J., Akalach, M., 2006. The effect of indoxacarb and five other insecticides on *Phytoseiulus persimilis* (Acari: Phytoseiidae), *Amblyseius fallacis* (Acari: Phytoseiidae) and nymphs of *Orius insidiosus* (Hemiptera: Anthocoridae). *Pest Manag. Sci.* 62, 334-339.

Bostanian, N.J., Akalach, M., 2004. The contact toxicity of indoxacarb and five other insecticides to *Orius insidiosus* (Hemiptera: Anthocoridae) and *Aphidius colemani* (Hymenoptera: Braconidae), beneficials used in the greenhouse industry. *Pest Manag. Sci.* 60, 1231-1236.

Bower, K.N., Berkett, L.P., Costante, J.F., 1995. Nontarget effect of a fungicide spray program on phytophagous and predacious mite populations in a scab-resistant apple orchard. *Environ. Entomol.* 24, 423-430.

Broadbent, A.B., Pree, D.J., 1984. Effects of diflubenzuron and BAY SIR 8514 on beneficial insects associated with peach. *Environ. Entomol.* 13, 133-136.

Brobyn, P.J., Clark, S.J., Wilding, N., 1988. The effect of fungus infection of *Metopolophium dirhodum* (Hom.: Aphididae) on the oviposition behavior of the aphid parasitoid *Aphidius rhopalosiphi* (Hym.: Aphidiidae). *Entomophaga* 33, 333-338.

Brooks, W.M. Host-parasitoid-pathogen interactions. In Beckage N.E, Thompson, S.N, Frederici, B.A editors. Pathogens (volume 2), Parasites and Pathogens of Insects. San Diego, CA: Academic Press; 1993; 231-272.

Burges, H.D., Hussey, H.W. Microbial control of insects and mites. London, UK: Academic Press; 1971.

Castagnoli, M., Liguori, M., Simoni, S., Duso, C., 2005a. Toxicity of some insecticides to *Tetranychus urticae*, *Neoseiulus californicus*, and *Tydeus californicus*. *BioControl* 50, 611-622.

Castagnoli, M., Nannelli, R., Simoni, S., 2005b. Effects of repeated applications of an azadirachtin-based product on the spider mite *Tetranychus urticae* and its phytoseiid predator *Neoseiulus californicus*. *Inter. Organ. Biol. Integr. Control/West Palaeartic Reg. Sec. (IOBC/WPRS) Bull.* 28, 51-54.

Castagnoli, M., Angeli, G., Liguori, M., Forti, D., Simoni, S., 2002. Side effects of botanical insecticides on predatory mite *Amblyseius andersoni* (Chant). *J. Pest Sci.* 75, 122-127.

Chauzat, M-P., Faucon, J-P., Martel, A-C., Lachaize, J., Cougoule, N., Aubert, M., 2006. A survey of pesticide residues in pollen loads collected by honey bees in France. *J. Econ. Entomol.* 99, 253-262.

Cherwonogrodzky, J.W., 1980. Microbial agents as insecticides. *Residue Rev.* 76, 73-96.

Clarkson, J.M., Charnley, A.K., 1996. New insights into the mechanisms of fungal pathogenesisin insects. *Trends Microbiol.* 4, 197-203.

Cloyd, R.A. Greenhouse pest management. Boca Raton, FL: CRC Press (Taylor & Francis Group); 2016.

Cloyd, R., 2016. Surface-level control. *Greenhouse Manag.* 36, 44-48.

Cloyd, R.A. Chapter 13: Insect and mite management in greenhouses. In Nelson P.V., editor. Greenhouse Operation and Management. 7th Edition. Upper Saddle River, NJ: Pearson Prentice Hall; 2012a; 391-441.

Cloyd, R.A. Indirect effects of pesticides on natural enemies. In Soundararajan R.P., editor. Pesticides—Advances in Chemical and Botanical Pesticides. Rijeka, Croatia: InTech; 2012b; 127-150.

Cloyd, R.A. Managing insect and mite pests. In Nau J, editor. Ball Redbook Vol. 2. 18th Edition. West Chicago, IL: Ball Publishing; 2011; 107-119.

Cloyd, R.A., 2006. Compatibility of insecticides with natural enemies to control pests of greenhouses and conservatories. *J. Entomol. Sci.* 41, 189-197.

Cloyd, R., 2004. Soaps and detergents: should they be used in interior plantscapes? *Southeastern Floric.* 14, 22-23.

Cloyd, R.A., Bethke, J. A., 2011. Impact of neonicotinoid insecticides on natural enemies in greenhouse and interiorscape environments. *Pest Manag. Sci.* 67, 3-9.

Cloyd, R.A., Timmons, N.R., Goebel, J.M., Kemp, K.E., 2009. Effect of pesticides on adult rove beetle [*Atheta coriaria* (Coleoptera: Staphylinidae) survival in growing medium. *J. Econ. Entomol.* 102, 1750-1758.

Cloyd, R.A., Galle, C.L., Keith, S., 2006. Compatibility of three miticides with the predatory mites *Neoseiulus californicus* McGregor and *Phytoseiulus persimilis* Athias-Henriot (Acari: Phytoseiidae). *HortScience* 41, 707-710.

Croft, B.A. Arthropod biological control agents and pesticides. New York, NY: John Wiley & Sons; 1990.

Croft, B.A., Brown, A.W.A., 1975. Responses of arthropod natural enemies to insecticides. *Annu. Rev. Entomol.* 20, 285-335.

Davidson, N.A., Dibble, J.E., Flint, M.L., Marer, P.J., Guye, A., 1991. Managing insects and mites with spray oils. *University of California Special Publ.* 3347, CA.

Desneux, N., Decourtye, A., Delpuech, J-M., 2007. The sublethal effects of pesticides on beneficial arthropods. *Ann. Rev. Entomol.* 52, 81-106.

Devine, G.J., Furlong, M.J., 2007. Insecticide use: contexts and ecological consequences. *Agric. Human Values* 24, 281-306.

Elzen, G.W., 2001. Lethal and sublethal effects of insecticide residues on *Orius insidiosus* (Hemiptera: Anthocoridae) and *Geocoris punctipes* (Hemiptera: Lygaeidae). *J. Econ. Entomol.* 94, 55-59.

Elzen, G.W. Sublethal effects of pesticides on beneficial parasitoids. In Jepson, P.C. editor. Pesticides and Non-Target Invertebrates. Wimborne, UK: Intercept; 1990; 129-150.

Elzen, G.W., O'Brien, P.J., Powell, J.E., 1989. Toxic and behavioral effects of selected insecticides on the parasitoid *Microplitis croceipes*. *Entomophaga* 34, 87-94.

Fiedler, Z., Sosnowska, D., 2014. Side effects of fungicides and insecticides on predatory mites, in laboratory conditions. *J. Plant Prot. Res.* 54, 349-353

Fiedler, Z., Sosnowska, D., 2007. Nematophagous fungus *Paecilomyces lilacinus* (Thom) Samson is also a biological agent for control of greenhouse insect and mite pests. *BioControl* 52, 547-558.

Fransen, J.J., van Lenteren, J.C.,1993. Host selection and survival of the parasitoid *Encarsia formosa* on greenhouse whitefly, *Trialeurodes vaporariorium*, in the presence of hosts infected with the fungus *Aschersonia aleyrodis*. *Entomol. Exp. Appl.* 69, 239-249.

Furlong, M.J., Pell, J.K. Interactions between entomopathogenic fungi and arthropod natural enemies. In Vega, F.E, and Blackwell, M. editors. Insect-Fungal Associations: Ecology and Evolution. New York, NY: Oxford University Press; 2005; 51-73.

Furlong, M.J., Pell, J.K., 2000. Conflicts between a fungal entomopathogen, *Zoophthora radicans*, and two larval parasitoids of the diamondback moth. *J. Invertebr. Pathol.* 76, 85-94.

Gentz, M.C., Murdoch, G., King, G.F., 2010. Tandem use of selective insecticides and natural enemies for effective reduced-risk pest management. *Biol. Control* 52, 208-215.

Gigon, V., Camps, C., Le Corff, J., 2016. Biological control of *Tetranychus urticae* by *Phytoseiulus macropilis* and *Macrolophus pygmaeus* in tomato production. *Exp. Appl. Acarol.* 68, 55-70.

Gillespie, A.T., Claydon, N., 1989. The use of entomogenous fungi for pest control and the role of toxins in pathogenesis. *Pest. Sci.* 27, 203-215.

Gonzalez, F., Tkaczuk, C., Dinu, M.M., Fielder, Z., Vidal, S., Zchori-Fein, E., Messelink, G.J., 2016. New opportunities for the integration of microorganisms into biological control systems in greenhouse crops. *J. Pest Sci.* 89, 295-311.

Gradish, A.E., Scott-Dupree, C.D., Shipp, L., Harris, C.R., Ferguson, G., 2011. Effect of reduced risk pesticides on greenhouse vegetable arthropod biological control agents. *Pest Manag. Sci.* 67, 82-86.

Hall, R.A., Burges, H.D., 1979. Control of aphids in glasshouses with the fungus, *Verticillium lecanii*. *Ann. Appl. Biol.* 93, 235-245.

Hassan, S.A., Van de Veire, M. Compatibility of pesticides with biological control agents. In Heinz, K.M., Van Driesche, R.G., Parrella, M.P. editors. Biocontrol in Protected Culture. Batavia, IL: Ball Publishing; 2004; 129-147.

Hayashi, H., 1996. Side effects of pesticides on *Encarsia formosa* Gahan. *Bull. Hiroshima Prefectural Agric. Res. Ctr*. 64, 33-43.

Haynes, K.F. 1988. Sublethal effects of neurotoxic insecticides on insect behavior. *Ann. Rev. Entomol*. 33, 149-168.

Heinz, K.M., Van Driesche, R.G., Parrella, M.P. editors. Biocontrol in protected culture. Batavia, IL: Ball Publishing; 2004.

Herrick, N.J., Cloyd, R.A., 2017. Direct and indirect effects of pesticides on the insidious flower bug (Hemiptera: Anthocoridae) under laboratory conditions. *J. Econ. Entomol*. (in press).

Hogendorp, B.K., Cloyd, R.A., 2013. Effect of potassium bicarbonate (MilStop®) and insecticides on the citrus mealybug, *Planococcus citri* (Risso), and the natural enemies *Leptomastix dactylopii* (Howard) and *Cryptolaemus montrouzieri* (Mulsant). *HortScience* 48, 1513-1517.

Holt, R.D., Hochberg, M.E., 1997. When is biological control evolutionarily stable (or is it)? *Ecology* 78, 1673-1683.

Isman, M.B., 2000. Plant essential oils for pest and disease management. *Crop Prot.* 19, 603-608.

Isman, M.B., Grieneisen, M.L., 2014. Botanical insecticide research: many publications, limited useful data. *Trends Plant Sci.* 19, 140-145.

Iwasa, T., Motoyama, N., Ambrose, J.T., Roe, R., Michael, M.R., 2004. Mechanism for the differential toxicity of neonicotinoid insecticides in the honey bee, *Apis mellifera*. *Crop Prot*. 23, 371-378.

Jacobson, R.J., Chandler, D., Fenlon, J., Russell, K.M., 2001. Compatibility of *Beauveria bassiana* (Balsamo) Vuillemin with *Amblyseius cucumeris* Oudemans (Acarina: Phytoseiidae) to control

Frankliniella occidentalis Pergande (Thysanoptera: Thripidae) on cucumber plants. *Biocontrol Sci. Technol.* 11, 391-400.

James, R.R., Lighthart, B., 1994. Susceptibility of the convergent lady beetle (Coleoptera: Coccinellidae) to four entomogenous fungi. *Environ. Entomol.* 23, 190-192.

James, R.R., Croft, B., Shaffer, B., Lighthart, B., 1998. Impact of temperature and humidity on host-pathogen interactions between *Beauveria bassiana* and a coccinellid. *Environ. Entomol.* 27, 1506-1513.

Jepsen, S.J., Rosenheim, J.A., Bench, M.E., 2007. The effect of sulfur on biological control of the grape leafhopper, *Erythroneura elegantula*, by the egg parasitoid *Anagrus erythroneurae*. *BioControl* 52, 721-732.

Kanzok, S.M., Jacobs-Lorena, M., 2006. Entomopathogenic fungi as biological insecticides to control malaria. *Trends Parasitol.* 22, 49-51.

Kim, D-S., Brooks, D.J., Riedl, H., 2006. Lethal and sublethal effects of abamectin, spinosad, methoxyfenozide and acetamiprid on the predaceous plant bug *Deraeocoris brevis* in the laboratory. *BioControl* 51, 465-484.

King, E.G., Bell, J.V., 1978. Interactions between a braconid, *Microplitis croceipes*, and a fungus *Nomuraea rileyi*, in laboratory-reared bollworm larvae. *J. Invertebr. Pathol.* 31, 337-340.

Kiselek, E.V., 1975. The effect of biopreparations on insect enemies. *Zashch. Rast.* 12, 23.

Kivett, J.M., Cloyd, R.A., Bello, N.M., 2015. Insecticide rotation programs with entomopathogenic organisms for suppression of western flower thrips (Thysanoptera: Thripidae) adult populations under greenhouse conditions. *J. Econ. Entomol.* 108, 1936-1946.

Koul, O., Walia, S., Dhaliwal, G.S., 2008. Essential oils as green pesticides: potential and constraints. *Biopestic. Int.* 4, 63-84.

Labbé, R.M., Gillespie, D.R., Cloutier, C., Brodeur, J., 2009. Compatibility of an entomopathogenic fungus with a predator and parasitoid in the biological control of greenhouse whitefly. *Biocontrol Sci. Technol.* 19, 429-446.

Labbé, R.M., Cloutier, C., Brodeur, J., 2006. Prey selection by *Dicyphus hesperus* of infected or parasitized greenhouse whitefly. *Biocontrol Sci. Technol.* 16, 485-494.

Lacey, L.A., Frutos, R., Kaya, H.K., Vail, P., 2001. Insect pathogens as biological control agents: do they have a future? *Biol. Control* 21, 230-248.

Lacey, L.A., Mesquita, A.L.M., Mercadier, G., Debire, R., Kazmer, D.J., Lecant, F., 1977. Acute and sublethal activity of the entomopathogenic fungus *Paecilomyces fumosoroseus* (Deuteromycotina: Hyphomycetes) on adult *Aphelinus asychis* (Hymenoptera: Aphelinidae). *Environ. Entomol.* 26, 1452-1460.

Livingston, J.M., Yearian, W.C., Young, S.Y., 1978. Effect of insecticides, fungicides and insecticide-fungicide combinations on developments of lepidopterous larval populations in soybean. *Environ. Entomol.* 7, 823-828.

Longley, M., Stark, J.D., 1996. Analytical techniques for quantifying direct, residual, and oral exposure of an insect parasitoid to an organophosphate insecticide. *Bull. Environ. Contam. Toxicol.* 57, 683-690.

Ludwig, S.W., Oetting, R.D., 2001. Susceptibility of natural enemies to infection by *Beauveria bassiana* and impact of insecticides on *Ipheseius degenerans* (Acari: Phytoseiidae). *J. Agric. Urban Entomol.* 18, 169-178.

Malezieux, S., Lapchin, L., Pralavorio, M., Moulin, J.C., Fournier, D., 1992. Toxicity of pesticide residues to a beneficial arthropod, *Phytoseiulus persimilis* (Acari: Phytoseiidae). *J. Econ. Entomol.* 85, 2077-2081.

Mar Fernández, M., Medina, P., Wanumen, A., Del Estal, P., Smagghe, G., Viñuela, E., 2017. Compatibility of sulfoxaflor and other modern pesticides with adults of the predatory mite *Amblyseius swirskii*. Residual contact and persistence studies. *BioControl* 62, 197-208.

Medina, P., Morales, J.J., Budia, F., Adan, A., Del Estal, P., Viñuela, E., 2007. Compatibility of endoparasitoid *Hyposoter didymator*

(Hymenoptera: Ichneumonidae) protected stages of five selected insecticides. *J. Econ. Entomol.* 100, 1789-1796.

Medina, P., Budia, F., del Estal, P., Viñuela, E., 2004. Influence of azadirachtin, a botanical insecticide, on *Chrysoperla carnea* (Stephens) reproduction: toxicity and ultrastructural approach. *J. Econ. Entomol.* 97, 43-50.

Medina, P., Smagghe, G., Budia, F., Tirry, L., Viñuela, E., 2003. Toxicity and absorption of azadirachtin, diflubenzuron, pyriproxyfen, and tebufenozide after topical application in predatory larvae of *Chrysoperla carnea* (Neuroptera: Chrysopidae). *Environ. Entomol.* 32, 196-203.

Mesquita, A.L.M., Lacey, L.A., 2001. Interactions among the entomopathogenic fungus, *Paecilomyces fumosoroseus* (Deuteromycotina: Hyphomycetes), the parasitod, *Aphelinus asychis* (Hymenoptera: Aphelinidae) and their aphid host. *Biol. Control* 22, 51-59.

Messelink, G.J., Bennison, J., Alomar, O., Ingegno, B.L., Tavella, L., Shipp, L., Palevsky, E., Wäckers, F.L., 2014. Approaches to conserving natural enemy populations in greenhouse crops: current methods and future prospects. *BioControl* 59, 377-393.

Meyling, N.V., Pell, J.K., 2006. Detection and avoidance of an entomopathogenic fungus by a generalist insect predators. *Ecol. Entomol.* 31, 162-171.

Mordue, A.J., Simmonds, M.S.J., Ley, S.V., Blaney, W.M., Mordue, W., Nasiruddin, M.,

Nisbet, A.J., 1998. Actions of azadirachtin, a plant allelochemical against insects. *Pestic. Sci.* 54, 277-284.

Moriarty, F., 1969. The sublethal effects of synthetic insecticides on insects. *Biol. Rev.* 44, 321-357.

Newsom, L.D., 1967. Consequences of insecticide use on nontarget organisms. *Ann. Rev. Entomol.* 12, 257-287.

Nicetic, O., Watson, D.M., Beattie, G.A.C., Meats, A., Zheng, J., 2001. Integrated pest management of two-spotted spider mite *Tetranychus urticae* on greenhouse roses using petroleum spray oil and the predatory mite *Phytoseiulus persimilis*. *Exp. Appl. Acarol*. 25, 37-53.

Oetting, R.D., Latimer, J.G., 1995. Effects of soaps, oils, and plant growth regulators (PGRs) on *Neoseiulus cucumeris* (Oudemans) and PGRs on *Orius insidiosus* (Say). *J. Agric. Entomol*. 12, 101-109.

Osborne, L.S., Bolckmans, K., Landa, Z., Peña, J. Kinds of natural enemies, In Heinz, K.M., Van Driesche, R.G., Parrella, M.P. editors. Biocontrol in Protected Culture. Batavia, IL: Ball Publishing; 2004; 95-127.

Osborne, L.S., Oetting, R.D., 1989. Biological control of pests attacking greenhouse grown ornamentals. *Florida Entomol*. 72, 408-413.

Osborne, L.S., Petitt, F.L., 1985. Insecticidal soap and the predatory mite, *Phytoseiulus persimilis* (Acari: Phytoseiidae), used in management of the of the twospotted spider mite (Acari: Tetranychidae) on greenhouse grown foliage plants. *J. Econ. Entomol*. 78, 687-691.

Otieno, J.A., Pallmann, P., Poehling, H-M., 2017. Additive and synergistic interactions amongst *Orius laevigatus* (Heteroptera: Anthocoridae), entomopathogens and azadirachtin for controlling western flower thrips (Thysanoptera: Thripidae). *BioControl* 62, 85-95.

Papaioannou-Souliotis, P., Markoyiannaki-Printziou, D., Zoaki-Malissiova, D., 2000. Side effects of Neemark (*Azadirachta indica* A. Juss) and two new vegetable oils formulations on *Tetranychus urticae* Koch and its predator *Phytoseiulus persimilis* Athias-Henriot. *Boll. Zool. Agr. Bachic. Ser. Il*. 32, 25-33.

Parrella, M.P. Arthropod fauna. In Stanhill, G., Zvi Enoch, H. editors. Ecosystems of the World 20. Greenhouse Ecosystems. New York, NY: Elsevier; 1999; 213-250.

Parrella, M.P., 1990. Biological control in ornamentals: status and perspectives. *Inter. Organ. Biol. Integr. Control/West Palaeartic Reg. Sec. (IOBC/WPRS) Bull*. 5, 161-168.

Parrella, M.P., Heinz, K.M., Nunney, L., 1992. Biological control through augmentative releases of natural enemies: a strategy whose time has come. *Am. Entomol*. 38, 172-179.

Pell, J.K., Pluke, R., Clark, S.J., Kenward, M.G., Alderson, P.G., 1997. Interactions between two aphid natural enemies, the entomopathogenic fungus *Erynia neoaphidis* Remaudiere & Hennebert (Zygomycetes: Entomophthorales) and the predatory beetle *Coccinella septempunctata* L (Coleoptera: Coccinellidae). *J. Invertebr. Pathol.* 69, 261-268.

Pilkington, L.J., Messelink, G., van Lenteren, J.C., Le Mottee, K., 2010. "Protected biological control"—Biological pest management in the greenhouse industry. *Biol. Control* 52, 216-220.

Pourian, H.R., Talaei-Hassanloui, R., Kosari, A.A., Ashouri. A., 2011. Effects of *Metarhizium anisopliae* on searching, feeding and predation by *Orius albidipennis* (Hem., Anthocoridae) on *Thrips tabaci* (Thy., Thripidae) larvae. *Biocontrol Sci. Technol.* 21, 15-21.

Powell, W., Wilding, N., Brobyn, P.J., Clark, S.J., 1986. Interference between parasitoids (Hym.: Aphidiidae) and fungi (Entomophthorales) attacking cereal aphids. *Entomophaga* 31, 293-302.

Puritch, G.S., Tonks, N., Downey, P., 1982. Effect of a commercial insecticidal soap on greenhouse whitefly (Hom: Aleyrod.) and its parasitoid, *Encarsia formosa* (Hym: Euloph.). *J. Entomol. Soc. Brit. Columbia* 79, 25-28.

Ripper, W.E., Greenslade, R.M., Lickerish, L.A., 1949. Combined chemical and biological control of insects by means of a systemic insecticide. *Nature* 163, 787-789.

Roger, C., Coderre, D., Vincent, D., 1994. Mortality and predation efficiency of *Coleomegilla maculata* lengi (Coleoptera: Coccinellidae) following pesticide applications. *J. Econ. Entomol.* 87, 583-588.

Rosenheim, J.A., Kaya, H.K., Ehler, L.E., Marois, J.J., Jaffee, B.A., 1995. Intra-guild predation among biological control agents: theory and evidence. *Biol. Control* 5, 303-335.

Rothwangl, K.B., Cloyd, R.A., Wiedenmann, R.N., 2004. Effects of insect growth regulators on citrus mealybug parasitoid *Leptomastix dactylopii* (Hymenoptera: Encyrtidae). *J. Econ. Entomol.* 97, 1239-1244.

Roubos, C.R., Rodriguez-Saona, C., Isaacs, R., 2014. Mitigating the effects of insecticides on arthropod biological control at field and landscape scales. *Biol. Control* 75, 28-38.

Roy, H.E., Pell, J.K., 2000. Interactions between entomopathogenic fungi and other natural enemies: implications for biological control. *Biocontrol Sci. Technol.* 10, 737-752.

Ruberson, J.R., Nemoto, H., Hirose, Y. Pesticides and conservation of natural enemies in pest management. In Barbosa P, editor. Conservation Biological Control. San Diego, CA: Academic Press; 1998; 207-220.

Saito, T., Brownbridge, M., 2016. Compatability of soil-dwelling predators and microbial agents and their efficacy in controlling soil-dwelling stages of western flower thrips *Frankliniella occidentalis*. *Biol. Control* 92, 92-100.

Saxena, R.C. Insecticides from neem. In Arnason, J.T., Philogene, B.J.R., Morand, P. editors. Insecticides of Plant Origin. ACS Symposium Series No. 387. Washington, D.C.: American Chemical Society; 1989; 110-135.

Schmutterer, H. 1990. Properties and potential of natural pesticides from the neem tree, *Azadirachta indica*. *Ann. Rev. Entomol.* 35, 271-297.

Schneider, M., Smagghe, G., Viñuela, E., 2004. Comparative effects of several insect growth regulators and spinosad on the different development stages of the endoparasitoid *Hyposoter didymator*. *Inter. Organ. Biol. Integr. Control/West Palaeartic Reg. Sec. (IOBC/WPRS) Bull.* 27, 13-19.

Seiedy, M., Saboori, A., Zahedi-Golpayegani, A., 2013. Olfactory response of *Phytoseiulus persimilis* (Acari: Phytoseiidae) to untreated and *Beauveria bassiana*-treated *Tetranychus urticae* (Acari: Tetranychidae). *Exp. Appl. Acarol.* 60, 219-227.

Shipp L., Kapongo, J.P., Park, H-H., Kevan, P., 2012. Effect of bee-vectored *Beauveria bassiana* on greenhouse beneficials under greenhouse cage conditions. *Biol. Control* 63, 135-142.

Shipp, J.L., Zhang, Y., Hunt, D.W.A., Ferguson, G., 2003. Influence of humidity and greenhouse microclimate on the efficacy of *Beauveria bassiana* (Balsamo) for control of greenhouse arthropod pests. *Environ. Entomol.* 32, 1154-1163.

Shipp, J.L., Boland, G.J., Shaw, L.A., 1991. Integrated pest management of disease and arthropod pests of greenhouse vegetable crops in Ontario: current status and future possibilities. *Can J. Plant Sci.* 71, 887-914.

Smith, S.F., Krischik, V.A., 2000. Effects of biorational pesticides on four cocinellid species (Coleoptera: Coccinellidae) having potential as biological control agents in interiorscapes. *J. Econ. Entomol.* 93, 732-736.

Sosnowska, D., Piatkowski, J., 1996. Efficacy of entomopathogenic fungus *Paecilomyces fumosoroseus* against whitefly (*Trialeurodes vaporariorum*) in greenhouse tomato cultures. *Inter. Organ. Biol. Integr. Control/West Palaeartic Reg. Sec. (IOBC/WPRS) Bull.* 19, 179-182.

Spollen, K.M., Isman, M.B., 1996. Acute and sublethal effects of neem insecticide on the commercial biological control agents *Phytoseiulus persimilis* and *Amblyseius cucumeris* (Acari: Phytoseiidae) and *Aphidoletes aphidimyza* (Diptera: Cecidomyiidae). *J. Econ. Entomol.* 89, 1379-1386.

Stara, J., Ourednickova, J., Kocourek, F., 2011. Laboratory evaluation of the side effects of insecticides on *Aphidius colemani* (Hymenoptera: Aphidiidae), *Aphidoletes aphidimyza* (Diptera: Cecidomyiidae), and *Neoseiulus cucumeris* (Acari: Phytoseidae). *J. Pest Sci.* 84, 25-31.

Stark, J.D., Vargas, R., Banks, J.E., 2007. Incorporating ecologically relevant measures of pesticide effect for estimating the compatibility of pesticides and biocontrol agents. *J. Econ. Entomol.* 100, 1027-1032.

Stark, J.D., Banks, J.E., 2003. Population-level effects of pesticides and other toxicants on arthropods. *Ann. Rev. Entomol.* 48, 505-519.

Stark, J.D., Jeppson, P.C., Mayer, D.F., 1995. Limitations to use of topical toxicity data for predictions of pesticide side effects in the field. *J. Econ. Entomol.* 88, 1081-1088.

Starnes, R.L., Liu, C.L., Marrone, P.G., 1993. History, use, and future of microbial insecticides. *Am. Entomol.* 39, 83-91.

Stevens, T.J., Kilmer, R.L., Glenn, S.J., 2000. An economic comparison of biological control and conventional control strategies for whiteflies (Homoptera: Aleyrodidae) in greenhouse poinsettias. *J. Econ. Entomol.* 93, 623-629.

Stevenson, J.H., Walters, J.H.H., 1983. Evaluation of pesticides for use with biological control. *Agr. Ecosyst. Environ.* 10, 201-215.

Sugiyama, K., Katayama, H., Saito, T., 2011. Effect of insecticides on the mortalities of three whitefly parasitoid species, *Eretmocerus mundus*, *Eretmocerus eremicus* and *Encarsia formosa* (Hymenotera: Aphelinidae). *Appl. Entomol. Zool.* 46, 311-317.

Tauber, M.J., Helgesen, R.G., 1978. Implementing biological control systems in commercial greenhouse crops. *Entomol. Soc. Amer. Bull.* 24, 424-426.

Theiling, K.M., Croft, B.A., 1988. Pesticide side-effects on arthropod natural enemies: a database summary. *Agr. Ecosyst. Environ.* 21, 191-218.

Tunaz, H., Uygun, N., 2004. Insect growth regulators for insect pest control. *Turk. J. Agric. For.* 28, 377-387.

van den Bosch, R., Stern, V.M., 1962. The integration of chemical and biological control of arthropod pests. *Annu. Rev. Entomol.* 7, 367-386.

Van Driesche, R.G., Heinz, K.M. An overview of biological control in protected culture. In Heinz, K., Van Driesche, R.G., Parrella, M.P. editors. Biocontrol in Protected Culture. Batavia, IL: Ball Publishing; 2004; 1-24.

van Lenteren, J.C. 2012. The state of commercial augmentative biological control: plenty of natural enemies, but a frustrating lack of uptake. *BioControl* 57, 1-20.

van Lenteren, J.C. 2000. A greenhouse without pesticides: fact or fantasy? *Crop Prot.* 19, 375-384.

van Lenteren, J.C., Woets, J., 1988. Biological and integrated pest control in greenhouses. *Ann. Rev. Entomol.* 33, 239-269.

Vestergaard, S., Gillespie, A., Butt, T., Schreiter, G., Eilenberg, J., 1995. Pathogenicity of the hyphomycete fungi *Verticillium lecanii* and

Metarhizium anisopliae to the western flower thrips, *Frankliniella occidentalis*. *Biocontrol Sci. Technol.* 5, 185-192.

Villanuéva-Jimenez, J.A., Hoy, M.A., 1998. Toxicity of pesticides to the citrus leaf miner and its parasitoid *Ageniaspis citricola* evaluated to assess their suitability for an IPM program in citrus nurseries. *BioControl* 43, 357-388.

Ware, G. W., Whitacre, D.M. The pesticide book (6th edition). Willoughby, OH: MeisterPro Information Resources; 2003.

Wawrzynski, R.P., Ascerno, M.E., McDonough, M.J., 2001. A survey of biological control users in Midwest greenhouse operations. *Am. Entomol.* 47, 228-234.

Williams, T., Valle, J., Vinuela, E., 2003. Is the naturally derived insecticide spinosad compatible with natural enemies? *Biocontrol Sci. Technol.* 13, 459-475.

Wright, D.J., Verkerk, R.H.J., 1995. Integration of chemical and biological control systems for arthropods: evaluation in a multitrophic context. *Pestic. Sci.* 44, 207-218.

Yardim, E.N., Edwards, C.A., 1998. The influence of chemical management of pests, diseases, and weeds on pest and predatory arthropods associated with tomatoes. *Agr. Ecosyst. Environ.* 70, 31-48.

Yu, S.J. The toxicology and biochemistry of insecticides. Boca Raton, FL: CRC Press (Taylor & Francis Group); 2008.

Biographical Sketch

Raymond A. Cloyd

Affiliation:

Department of Entomology, Kansas State University, Manhattan, KS US

Education:

Doctor of Philosophy Degree (Ph.D) in Entomology; Purdue University, West Lafayette, IN. March 1999.

Dissertation: Effects of plant architecture on the attack rate of *Leptomastix dactylopii* (Howard) (Hymenoptera: Encyrtidae), a parasitoid of the citrus mealybug, *Planococcus citri* (Risso) (Homoptera: Pseudococcidae).

Master of Science Degree in Entomology; Purdue University, West Lafayette, IN. June 1995. Thesis: Evaluation of trichome density levels and bean leaf beetle, *Cerotoma trifurcata* (Forster), feeding preference on soybean pods.

Bachelor of Science Degree in Ornamental Horticulture; California Polytechnic State University—San Luis Obispo, CA. June 1990. Senior Project: Fungicide efficacy on *Pythium ultimum* of bedding plant seedlings. Minor Degree in Plant Protection/Pest Management.

Associate of Science Degree in Ornamental Horticulture; Monterey Peninsula College, Monterey, CA. June 1985.

Business Address:

Department of Entomology, Kansas State University, 123 Waters Hall, Manhattan, KS 66506 USA

Research and Professional Experience:

Dr. Raymond A. Cloyd is an extension specialist (70% of appointment) in horticultural entomology/plant protection and serves as the state entomology extension leader for the Department of Entomology at Kansas State University (Manhattan, KS). His responsibilities include: greenhouses, nurseries, landscapes, turfgrass, conservatories, interiorscapes, Christmas trees, vegetable and fruits, and beekeepers. Dr. Cloyd's clientele groups include homeowners, master gardeners, commercial producers, and professional and commercial operators. He has been involved in 43 extension publications that were published in 2010 through 2016. Dr. Cloyd is responsible for all the Master Gardener Training Programs associated with entomology and provides training for the Kansas Arborists Association Arborists Training Course every year. He has given 348 extension-related presentations both regionally and nationally, and has written 152 trade journal articles from 2010 to 2016.

Dr. Raymond A. Cloyd's research program (30% of appointment) compliments his extension program where he conducts applied or practical research projects that will help his clientele to effectively deal with insect and mite pests of horticultural crops. He has published 31 peer-reviewed papers in various journals, from 2010 to 2016, associated with plant protection. Dr. Cloyd's research program focuses on three key insect pests: western flower thrips (*Frankliniella occidentalis*), citrus mealybug (*Planococcus citri*), and fungus gnat (*Bradysia* nr. sp. *coprophila*). These three are considered to be major insect pests of horticultural cropping systems. One of his publications [Cloyd, R. A., K. A. Marley, R. A. Larson, and B. Arieli. 2010. Bounce® fabric softener dryer sheets repel fungus gnat, *Bradysia* sp. nr. *coprophila* (Diptera: Sciaridae), adults. HortScience 45(12): 1830-1833] was the number one downloaded article from December 2010 to August 2011. In fact, according to Mike Neff (Executive Director: American Society of Horticultural Science) "1100 is the most accesses we've ever had for an article in any of the journals and since publication of the article, it has been downloaded 1,920 times. That's a lot!" Dr. Cloyd had 21 grant proposals funded, from 2010 to 2016, for a total of $910,503.00. In addition to the funds obtained from competitive grants, he has acquired $246,250.00 from industry to conduct efficacy trials with existing products or evaluate new products prior to commercial availability. Dr. Cloyd's research program has been instrumental in developing rotation programs for use against the western flower thrips, which are designed to reduce the potential for insecticide resistance developing in western flower thrips populations. Another important aspect of his research program has been associated with understanding the impact of pesticides on biological control agents (e.g., parasitoids and predators) so that greenhouse producers can integrate pesticides with biological control in order to more effectively deal with insect and mite pests.

Professional Appointments:

Assistant Professor/Associate Professor: Extension Specialist in Ornamental Entomology/Integrated Pest Management; University of Illinois, Department of Natural Resources and Environmental Sciences, Urbana, IL. 1999-2006. Appointment: Extension (70%) and Research (30%).

Responsibilities: Extension and research associated with pest management of greenhouses, nurseries, landscapes, turfgrass, conservatories, and interiorscapes. Extension specialist in ornamental entomology for the state of Illinois. Major clientele: homeowners, master gardeners, and professional and commercial operators.

Honors:

University of Illinois Extension Outstanding/Innovative Program Team Award ("Insect Identification Series") 2001.

Southern Illinois Bedding Plant School Outstanding Service Award 2003.

College of Agricultural, Consumer, and Environmental Sciences Faculty Award for Excellence in Extension 2003.

Indiana Flower Growers Association Appreciation Award 2003.

Visionary Leadership Award, Epsilon Sigma Phi Extension Fraternity 2003.

American Society of Horticultural Science, Outstanding Extension Publication Award (Outstanding Primarily Visual Award) 2003.

University of Illinois Excellence in Extension Campus Award for Less than 10 Years of Service Award 2004.

Entomological Society of America Outstanding Service Award, Arthropod Management Test Editorial Board (Chairperson and Member) 2005.

Early Career Extension Award, Epsilon Sigma Phi Extension Fraternity (Illinois Alpha Nu Chapter) 2006.

North Central Region, Epsilon Sigma Phi Early Career Extension Award 2007.

Epsilon Sigma Phi, Alpha Rho State Early Career Extension Award 2008.

Extension Length of Service Award—10 years: Epsilon Sigma Phi Alpha Rho Chapter 2009.

Kansas State Research and Extension 10 Years of Service Award; Kansas State University 2009.

Ohio Florists' Association Bulletin Author of the Year, Ohio Florists' Association Award 2010.

American Rose Society Award of Merit Bulletin/Newsletter Contest (Article: Dormant oils—Everything you ever wanted to know. The Nashville Rose Leaf) 2010.

American Rose Society Award of Merit Bulletin/Newsletter Contest (Article: Defusing misconceptions about thrips. The Nashville Rose Leaf) 2010.

American Rose Society Award of Merit Bulletin/Newsletter Contest (Article: What rose growers should know about colony collapse disorder. The Nashville Rose Leaf) 2010.

Entomological Society of America North Central Branch Award of Excellence in Integrated Pest Management 2010.

American Society for Horticultural Science Outstanding Extension Publication Award (Outstanding Leaflet Award; Resistance Management: Resistance, Mode of Action, and Pesticide Rotation) 2010.

American Rose Society Award of Merit Bulletin/Newsletter Contest (Article: Pesticide resistance. The Nashville Rose Leaf) 2011.

American Rose Society Award of Merit Bulletin/Newsletter Contest (Article: Imidacloprid (Merit®). The Nashville Rose Leaf) 2011.

Society of American Florists Alex Laurie Award for Research and Extension 2011.

American Rose Society Award of Merit Bulletin/Newsletter Contest (Article: Rose rosette disease. The Nashville Rose Leaf) 2012.

American Rose Society Award of Merit Bulletin/Newsletter Contest (Article: What are pesticide metabolites. The Nashville Rose Leaf) 2012.

American Society for Horticultural Sciences Outstanding Extension Educator Award 2012.

American Rose Society Award of Merit Bulletin/Newsletter Contest (Article: Japanese beetles: What can you do? The Nashville Rose Leaf) 2013.

American Rose Society Award of Merit Bulletin/Newsletter Contest (Article: Two spotted spider mites on roses. The Nashville Rose Leaf) 2013.

Epsilon Sigma Phi Alpha Rho Chapter Mid-Career Extension Award 2013.

American Rose Society Award of Merit Bulletin/Newsletter Contest (Article: METI Miticides: What are They? The Nashville Rose Leaf) 2014.

American Rose Society Award of Merit Bulletin/Newsletter Contest (Article: Rose aphid—Beware of this 'sucker'. The Nashville Rose Leaf) 2014.

American Rose Society Award of Merit Bulletin/Newsletter Contest (Article: Soaps and detergents: Should they be used on roses? The Nashville Rose Leaf) 2014.

American Rose Society Award of Merit Bulletin/Newsletter Contest (Article: What is insecticide hormoligosis? The Nashville Rose Leaf) 2014.

American Society for Horticultural Science Extension Division; Outstanding Education Materials Award (Outstanding Leaflet Award: Rose Rosette Disease) 2014.

K-State Research and Extension 2014 Team Award; Extension Integrated Pest Management 2014.

Alpha Rho Chapter, Epsilon Sigma Phi; National Honor Extension Fraternity; Tenure Award (15 years: Cooperative Extension Service, Kansas State University, Manhattan, KS) 2014.

American Rose Society Award of Merit Bulletin/Newsletter Contest (Article: Translaminar pesticides: Factors that may impact effectiveness. The Nashville Rose Leaf) 2015.

American Rose Society Award of Merit Bulletin/Newsletter Contest (Article: What are adjuvants? The Nashville Rose Leaf) 2015.

American Society for Horticultural Science Extension Division; Outstanding Education Materials Award (Outstanding Leaflet Award: Spotted Wing Drosophila) 2015.

National Association of County Agricultural Agents (NACAA) Communications Contest [Publication: "Commercially available biological control agents for common greenhouse insect pests." Authors: Heidi Wollaeger and David Smitley (Michigan State University), and Raymond Cloyd (Kansas State University)] 2016.

Publications from the Last 3 Years (2016-2014):

Cloyd, R. A. 2016. Western flower thrips (Thysanoptera: Thripidae) and insecticide resistance: an overview and strategies to mitigate insecticide resistance development. J. Entomol. Sci. 51: 257-273.

Kivett, J. M., R. A. Cloyd, and N. M. Bello. 2016. Evaluation of entomopathogenic fungi against the western flower thrips (Thysanoptera: Thripidae) under laboratory conditions. J. Entomol. Sci. 51: 274-291.

Herrick, N. J., and R. A. Cloyd. 2016. Western flower thrips, *Frankliniella occidentalis* (Thysanoptera: Thripidae): evaluation of potential attraction to vanilla extract under laboratory and greenhouse conditions. HortScience 51: 525-529.

Cloyd, R. A. 2015. Western flower thrips management in greenhouse production systems in the 21st century: alternative strategies need be considered. Acta Hort. 1104: 381-394.

Everman, E. R., R. A. Cloyd, C. Copland, and T. J. Morgan. 2015. First report of spotted wing drosophila, *Drosophila suzukii* Matsumura (Diptera: Drosophilidae) in Kansas. J. Kansas Entomol. Soc.88: 128-133.

Echegaray, E. A., R. A. Cloyd, and J. R. Nechols. 2015. Rove beetle (Coleoptera: Staphylinidae) predation on *Bradysia* sp. nr. *coprophila* (Diptera: Sciaridae). J. Entomol. Sci. 50: 225-237.

Kivett, J. M., R. A. Cloyd, and N. M. Bello. 2015. Insecticide rotation programs with entomopathogenic organisms for suppression of western flower thrips (Thysanoptera: Thripidae) adult populations under greenhouse conditions. J. Econ. Entomol. 108: 1936-1946.

Cloyd, R. A. 2015. Ecology of fungus gnats (*Bradysia* spp.) in greenhouse production systems associated with disease-interactions and alternative management strategies. Insects 6: 325-332.

Cloyd, R. A., and A. L. Raudenbush. 2014. Efficacy of binary pesticide mixtures against western flower thrips. HortTechnology 24(4): 449-456.

Raudenbush, A. L., R. A. Cloyd, and E. R. Echegaray. 2014. Effect of a physical barrier on adult emergence and egg survival associated with the fungus gnat, *Bradysia* sp. nr. *coprophila* (Diptera: Sciaridae), under laboratory conditions. HortScience 49(7): 905-910.

In: Biological Control
Editor: Lewis Davenport
ISBN: 978-1-53612-416-3

Chapter 4

BIOLOGICAL CONTROL OF FUNGAL DISEASES BY ANTAGONISTIC MICROORGANISMS

Ivana Potočnik *and Svetlana Milijašević-Marčić***
Laboratory of Applied Phytopathology, Institute of Pesticides and Environmental Protection, Belgrade-Zemun, Serbia

ABSTRACT

Worldwide plant disease control usually relies on the use of fungicides. However, pathogen resistance to fungicides, developing after frequent treatments, and host sensitivity to fungicides have been recognized as serious problems. Disease control with only a few or no chemicals available at all is a major challenge for growers in the 21st century. Considering pathogen resistance evolution and harmful impact on the environment and human health, special attention has been focused on biofungicides based on antagonistic microorganisms. Previous studies have indicated that various biological agents appear to be suitable, eco-

* Corresponding Author Email: ivana.potocnik@pesting.org.

friendly alternatives as they have demonstrated strong antifungal effects against pathogenic fungi on cultivated plants and mushrooms. Microbiological products are usually based on antagonistic microorganisms, such as bacteria (*Bacillus* spp., *Lactobacillus* spp. and *Pseudomonas* spp.), fungi (*Pythium* spp. and *Trichoderma* spp.) and actinomycetes (*Streptomyces* spp., etc.). The use of microbial inoculants registered for biological control applications is the result of their ability to antagonize the pathogen by multiple modes of action and also to effectively colonize the rhizosphere or phyloshere. Strong activities of antagonistic species are due to their production of antibiotics, volatile compounds, lytic enzymes, and plant growth-promotion effects by phytohormones. The efficacy of antagonistic organisms depends on many abiotic and biotic factors, climatic conditions and interactions with other soil-inhabiting organisms. The introduction of biofungicides has created new possibilities for crop protection with reduced application of chemicals.

Keywords: biofungicides, antagonistic microorganisms, fungi, plants, mushrooms

INTRODUCTION

The terms "biological control" or "biocontrol" in plant pathology mostly refer to the use of microbial antagonists to suppress diseases, and the organism that suppresses the pathogen is referred to as the biological control agent (BCA) (Pal and McSpadden Gardener, 2006). In a more limited sense, biological control refers to the suppression of a single pathogen (or pest) by a single antagonist in a single cropping system. As a result of BCA applications, the incidence and severity of diseases have decreased (Haas and de Fago 2005; Pal and McSpadden Gardener, 2006).

Antagonistic microorganisms have been reported to possess significant antagonistic activities against a range of fungal plant pathogens (Hasan et al., 2013). These antagonists include beneficial bacteria (*Bacillus subtilis, B. amyloliquefaciens*, *Pseudomonas fluorescens,* etc.), fungi (*Trichoderma virens, T. harzianum, Gliocadium* spp., etc.) and actinomycetes (*Streptomyces* spp., etc.). Microbial antagonists occupy the same ecological niches as the target plant pathogens, interacting with them directly or indirectly. These rhizosphere and endophyte microorganisms that are used for plant disease biological control have a wide range of modes of action. The mechanisms of direct interaction include parasitism, competition for space, water or nutrients, and production of antibiotics and siderophores or synthesis of hydrolytic enzymes (chitinases, glucanases, proteases, and lipases) with different effects on target pathogens. One of the most closely studied modes of action of bacterial BCAs is the antagonism induced by different compounds with antifungal properties (Haas and Keel, 2003). Most microorganisms produce and secrete one or more compounds with antibiotic activity (Raaijmakers et al., 2002), such as the biocontrol strain *Pseudomonas fluorescens* Pf-5, which produces the antibiotics pyrrolnitrin, pyoluteorin and 2,4-diacetylphloroglucinol (Loper et al., 2007). Additionally, antifungal activity has been reported for many different lipopeptides (Raajmakers et al., 2006). Dimkić et al. (2013) revealed that strains SS-13.1 and SS-12.6, identified as *Bacillus amyloliquefaciens,* had potential for production of iturin A, bacillomycin and surfactin. Also, Koumoutsi et al. (2004) found that the strain FZB42 of *B. amyloliquefaciens* produces various antifungal lipopeptides. On the other hand, some strains of rhizosphere bacteria, called plant growth-promoting rhizobacteria (PGPR), stimulate plant growth by directly affecting plant metabolism and/or the availability of nutrients. Other PGPR strains promote plant growth indirectly by suppressing soil-borne pathogens through production of bacterial metabolites (antibiotics, iron chelators, cell wall degrading enzymes, and hydrogen cyanide) that adversely affect the pathogen (Kloepper et al., 2004; Milijašević-Marčić and Todorović, 2017). Biocontrol fungi can use mechanisms such as induced systemic resistance (ISR) (Fuchs et al., 1997; Duijff et al., 1998), predation and parasitism (Harman et al., 2004; Bolwerk, 2005) and competition for niches and nutrients (Bolwerk et al., 2005).

Over the years, many bacterial isolates have been evaluated as potential biocontrol agents against plant pathogenic fungi. However, only a few of them were ultimately successful after evaluation in the field or *in planta* trials. The reasons for this failure could be found in a lack of appropriate screening procedures to select the most suitable microorganisms for disease control in diverse environments (Pliego et al., 2011). In addition, the approach may turn out inappropriate because it cannot anticipate the variety of host-antagonist-pathogen interactions before selecting those biocontrol agents that provide disease control through mechanisms such as root colonization, induction of systemic resistance or niche competition (Bakker et al,. 2003; Lugtenberg and Kamilova, 2009).

Future outlook of biocontrol of plant diseases seems promising due to a growing demand for biologically-based pest management practices. Recent surveys of both conventional and organic growing practices on farms indicated an interest in using biocontrol products, suggesting that the market potential of biocontrol products will be increasing in coming years (Pal and McSpadden Gardener, 2006). Although research has lead over the past years to development of a small commercial sector that produces a number of biocontrol products, the major challenge is yet to develop a formulation and application method that can be implemented on a commercial scale, while being at the same time effective, reliable, consistent, economically feasible, and with a wider spectrum (Hasan et al., 2013).

In this chapter, we will discuss screening methods for candidate biocontrol agents, application methods and challenges associated with biological control of fungal pathogens by antagonistic microorganisms.

The Screening Methods for Candidate Biocontrol Agents

The first step in finding a biocontrol candidate is to build up a collection of bacterial isolates from an appropriate environment. Selection of isolation sites depends on experimental system and type of the pathogen aimed to control. A basic principle that should be considered for a successful biological control programme is good adaptation of a BCA to the local environmental conditions in which it is expected to work. Therefore, some authors have proposed isolation of bacterial antagonists from the rhizosphere and phylosphere of pathogen host plants (Romero et al., 2004; Milijašević-Marčić et al., 2017). The choice of a screening site is also based on areas with high disease pressure and lack of disease symptoms (Pliego et al., 2011). In addition, disease-suppressive soils are considered as an appropriate site for selecting naturally-occurring BCAs for soil-borne diseases (Haas & de Fago, 2005).

Screening strategies for biocontrol of plant fungal pathogens have a number of technical and conceptual requirements. Methods depend on the type of biocontrol strategy (preventive or curative), environmental conditions, pathogen biology and other factors. Pliego et al. (2011) summarized the types of screening methods found in the relevant literature as: *in vitro* assays, assays involving plants, screening for plant-growth promoting rhizobacteria, screening for colonization, screening for induced resistance, screening by plant performance, and screening of endophytes and molecular techniques of screening.

In vitro screening methods include plate assays with only one microorganism (aiming at lytic enzymes or siderophores production) or with two different microorganisms (mainly searching for antagonistic or parasitic relationships). Search for enzyme-producing microorganisms is a rapid and simple procedure performed prior to interaction with the pathogen and plant. The main lytic enzymes produced by bacterial BCAs are chitinases, glucanases and proteases, although chitinolysis is a common trait in bacteria that exhibit antifungal activity (de Boer et al., 2005; Hoster et al., 2005). For screening based on nonantibiotic substances, such as siderophore production, specific media should be used to select bacterial BCAs belonging to the genera *Pseudomonas, Bacillus* and *Kocuria* that are active against many fungal soil-borne phytopathogens, such as *Fusarium oxysporum*, *Pyricularia oryzae*, and *Sclerotium* spp. (Chaiharn et al., 2009).

Screening for antagonism against fungal pathogens is one of the most straightforward methods used for the selection of bacterial biocontrol agents and it has been proved to be a valid strategy at least for soil-borne pathogenic microorganisms. Antagonism, as a mode of action, could be considered as a method of inhibiting phytopathogenic fungi through secretion of substances that interfere with the life cycle of the target microorganism. Antagonistic bacterial-fungal interactions are typically assessed *in vitro* in terms of an unoccupied "inhibition zone" between a bacterial colony and fungal hyphae cocultured on an agar plate. Two-component screening (e.g., dual cultures of a candidate antagonist and a pathogen on agar) is exclusively related to interaction studies, and potential antagonists are typically ranked according to their ability to inhibit the growth of the pathogen expressed by an inhibition zone. The antibiotic producing strains have been studied for their antagonism achieved that way, and such antibiotics are known to be active against fungi *in vivo*. The production of these antagonistic substances sometimes correlates very well with the biocontrol ability of these bacteria, and the dual culture method has shown to be suitable for their screening. However, *in vitro* dual culture screening is mostly related to interaction studies and some authors have found that production of antibiotics *in vitro* is not consistent with their production *in vivo*. One reason for this lack of consistency could be that production of antibiotics depends on specific nutrients that are present in the media (Pliego et al., 2011).

As reported before (Walker et al., 1998; Yoshida et al., 2001) the technique of dual culture on agar plates proved to be an efficient and easy way to select antagonistic bacteria from a random group of bacterial isolates based on their fungal growth inhibition capability. There are several types of dual culture tests on agar plates which depend on the pathogen and antagonist involved. In a bacteria versus fungi system, an antagonist is placed/streaked on one side of the plate and a pathogenic fungi on the other side (Stanojević et al., 2016; Milijašević-Marčić et al., 2017), and then inhibition is measured and inhibition percentage calculated. In the agar diffusion method, an antagonist is added into a well and a pathogen into the media (radial growth inhibition assay). Yet another method includes point inoculation with bacterial isolates and subsequent inoculation with pathogen spore suspension using a test tube atomizer (Berendsen et al., 2012). Additionally, volatile metabolites could be assessed with a paired Petri dish technique where a bacterial broth suspension is spread on one plate and a pathogen placed in the center of another plate and then one of them is placed upside down over another one avoiding physical contact between them (Solanki et al., 2013). Moreover, the dual culture technique can also be used for a diffusible metabolite assay. In that method, bacterial antagonists are grown in appropriate sterilized nutrient broth in the incubator shaker (180 rpm at 28°C) and a cell-free supernatant is obtained by centrifugation of the culture filtrates of the respective organism at 8000g for 20 minutes at 4°C. After centrifugation, the culture filtrates need to be filtered through sterile syringe filters of 2 mm pore size and applied into the wells made in agar plates inoculated with a pathogen (Berić et al., 2012; Solanki et al., 2013; Milijašević-Marčić et al., 2016).

In a fungi vs fungi system, antagonism is investigated by placing discs of a potential antagonistic fungi and a pathogen 2 cm apart from each other in a Petri dish, and then antagonistic potential is assessed by measuring colony growth on both sides, i.e., towards and opposing each other from the loci. The parameters that could be used for assessments of colony interaction are the degree of inhibition or intermingled zone between the two colonies. The percent inhibition of radial growth could be calculated by using the equation put forward by Fokkema (1973) (Basumatary et al., 2015).

For actinobacteria *vs* fungi testing by dual culture technique, a fungal isolate is placed 4 cm from an actinobacterial streak line where actinobacterial isolate was previously grown by streak-plate method on specific media for 5 to 7 days. The size of any apparent inhibition zone in the interaction area of both microorganisms is measured (approximately a week after incubation) (Nakaew et al., 2015).

In addition, *in vitro* screening for antagonists by dual cultures is an easy and inexpensive method that permits large-scale screening of several strains of microorganisms. If the goal is to select microorganisms with high capabilities of natural metabolite production and to develop these natural products for commercial applications, prescreening for antibiosis may be an appropriate method (Pliego et al., 2011). However, some authors have suggested that identification of effective antagonist strains is only the first step in developing effective biological control (Larkin and Fravel, 1998). For biocontrol to be implemented on a practical level, the antagonists have to be ecologically able to survive, become established, and function under particular conditions of an appropriate ecosystem (Pliego et al., 2011).

In the absence of plants, selection for antagonism alone will not provide information concerning the ability of a microorganism to colonize and protect roots and seeds. In addition, even though antibiosis or mycoparasitism have been shown to occur, it is often the competition for nutrients and ability to compete against other organisms in the rhizosphere, spermosphere, and other areas that are the essential characteristics of successful biocontrol organisms (Lugtenberg and Kamilova, 2009). Therefore, *in planta* assays were developed to enable studies of the host-antagonist-pathogen interactions and selection of biocontrol agents that provide disease control through mechanisms such as induced resistance, root colonization, plant growth promotion or niche competition (Bakker et al,. 2003; Lugtenberg and Kamilova, 2009). Kuiper et al. (2001) described a method to select enhanced grass root tip colonizing bacteria. In this method a mixture of rhizosphere bacteria were applied to a sterile seedling. After a period of plant growth, the bacteria that have reached the root tip were isolated. These were subsequently used to inoculate a new sterile seedling, which was again allowed to grow. After three of these enrichment cycles, excellent competitive root tip colonizers were obtained (Kuiper et al., 2001). Kamilova et al. (2005) used the same method to select enhanced tomato and cucumber root tip colonizers. Based on their observation that not only biocontrol fungi exist but perhaps also biocontrol bacteria, which act through the mechanism 'competition for niches and nutrients', the authors screened some selected enhanced root tip colonizers for their ability to control the tomato foot and root rot, a disease caused by the fungus *Fusarium oxysporum* f. sp. *radicis-lycopersici*.

In addition to searching for candidate BCAs, *in planta* assays can be used after *in vitro* screening to confirm their efficacy against fungal pathogens in plants or detached plant parts. In a study by Dimkić et al. (2013), surface sterilized apple fruits were wounded and co-inoculated with extracts of *Bacillus* spp. strains and postharvest pathogens. Isolate SS-12.6 strongly inhibited the growth of *Colletotrichum acutatum* and *C. gloeosporioides* (PGI 100%), *Monilinia fructigena* (PGI 91.5%), *Alternaria alternata* (PGI 90.4%), *Fusarium solani* (PGI 82.9%), *Botryosphaeria obtusa* (PGI 81.6%), *Fusarium oxysporum*. (PGI 67.1%), *Penicillium expansum* (PGI 77.1%), *Mucor* sp. (PGI 75.2%), and *Aspergillus flavus* (PGI 74.8%). *In planta* assays that followed *in vitro* screening, revealed a significant reduction in necrosis caused by several postharvest fungal pathogens, such as *M. fructigena* (66.6%), *P. expansum* (57.7%), *F. oxysporum* (52.4%), *F. solani* (47.4%), and *Mucor sp*. (46%), compared to a positive control (Dimkić et al., 2013). Berendsen et al. (2012) conducted an assay for antagonism of several *Pseudomonas* spp. isolates on mushroom caps (in a mushroom growing room and on detached mushroom saps) after *in vitro* tests against the causal agent of dry bubble disease by adding mixtures of bacterial cells and conidia of mushroom pathogenic fungi *Lecanicillium fungicola* f. sp. *fungicola.*

A direct positive influence on plant growth, and indirect stimulation of plant health by applying root-associated bacteria is also an important criterion for BCAs. Several plate tests have been developed to test plant growth-promotion using plants with small seeds, such as a microplate assay with strawberry seedlings (Berg et al., 2001). These plant growth promotion assays in microplates are simpler *in planta* tests that can imitate and replace a whole plant system considering time consumption, plant material, and growth facilities. In addition, it has the advantages of allowing many repetitions and high throughput screening for a large number of bacterial isolates. On the basis of *in vitro* testing, an assessment system should be developed to select the most efficient BCAs for greenhouse trials which should be followed by field trials under different climatic conditions and diverse soil qualities (Berg, 2007).

It is well-known that microbial antagonists occupy the same ecological niches as the target plant pathogen interacting with it directly or indirectly. Interaction of some rhizosphere microorganisms with the plant roots can create plants resistant to some pathogenic bacteria, fungi and viruses. This phenomenon is called ‘induced systemic resistance’ (ISR). ISR is dependent on jasmonic acid and ethylene signalling in the plant, and it can be induced by non-pathogenic bacteria in the soil (Kloepper et al., 2004).

ISR differs from SAR (systemic acquired resistance), which is associated with an increase in accumulation of salicylic acid (SA), a compound which is frequently produced after pathogen infection, and with an increased expression of several pathogenesis-related (PR) proteins (Pal and McSpadden Gardener, 2006). To select bacterial BCAs with this trait, an easy screening method based on PGPR could be developed. Using a plate assay, the increase in growth of root and/or aerial plant (such as tobacco, tomato, lettuce, arabidopsis, etc.) can be measured after exposure to a potential BCA or its exudates incorporated to the media (van Loon, 2007). Such screening methods could indicate that multiple defense mechanisms in plants can be activated by, and may be effective against, diseases caused by a broad range of fungi, bacteria and viruses. A second round of experiments with inclusion of the pathogen could lead to the selection of potential bacterial BCAs. Therefore, only screening methods that include plants can be used for testing the phenomenon of induced resistance in the evaluation of potential BCAs (Shoresh et al., 2010).

In addition, plant endophyte microorganisms can also influence plant growth. The genera *Bacillus, Pseudomonas, Serratia*, *Arthrobacter, Micrococcus* and *Curtobacterium* include endophytic representatives (Aravind et al., 2009; 2010). Isolation of a candidate plant root endophyte for a BCA is usually performed by isolating microorganisms associated with washed and surface-sterilized roots. After obtaining bacterial candidates, the next step is a search for antagonistic endophytes by dual-plate assays (Aravind et al., 2009).

Finally, there are molecular techniques of screening for BCAs. Considering that selection and evaluation of microbial strains for their antifungal activity under specific environmental conditions is time consuming and does not allow testing of a large number of strains, molecular techniques which rely on a genetic screening approach have been developed recently (Pliego et al., 2011). For example, Giacomodonato et al. (2001) developed a PCR-based method to search for peptide-producing microorganisms, which resulted in the selection of *Bacillus* strains with antifungal activity against *Sclerotinia sclerotiorum*..

There are two culture-independent approaches that can be followed in order to rapidly reveal functionally important microbes such as BCAs. The first approach is to use genetic markers for a functionally important activity such as antibiosis. By isolating a diverse set of genetic variants, strains or subspecies with various capacities to colonize plant roots and/or suppress pathogens can be identified (Pliego et al., 2011). In the absence of knowledge about the mechanisms involved in biocontrol, PCR-based suppressive-subtractive hybridisation can be used to identify new markers (Leveau et al. 2006). The second approach is based on molecular profiling of microbial population structure where ribosomal gene sequences are targeted, amplified from the rhizosphere environment, and analysed. Low-cost, low-resolution techniques, such as terminal restriction fragment (TRF) length polymorphism (T-RFLP) analyses, provide a cost-effective approach to find populations that consistently contribute to suppression across environments. T-RFLP analyses compare the bacterial community structure in soils differing in their disease-suppressive capacities, revealing the positive association of multiple bacterial populations (marked with different TRFs) with disease suppression (Benitez and McSpadden Gardener 2009). Evolution of molecular tools has recently enabled the development of new screening strategies, such as the development of sequence based T-RFLP-derived molecular markers to direct the identification and isolation of novel bacteria (Benitez and McSpadden Gardener 2009; Pliego et al., 2011).

Applications of Microbial Biofungicides

The use of beneficial microorganisms is considered one of the most promising alternatives to synthetic fungicides. Biopesticides have several significant properties: high specificity against target pathogens, easy degradation and low mass production cost. The modes of action of antagonistic microorganisms include competition with pathogens for space and nutrients, production of antibiotics and cell-wall degrading enzymes, or reduction of pathogen population by hyperparasitism (Milijašević-Marčić et al., 2017). Although fungitoxicity of some biopesticides is not strong, they can be used as supplements to commercial formulations, which will minimaze the quantity of fungicides used (Todorović et al., 2012).

Antagonistic Fungi

***Trichoderma* spp.** Fungi in the genus *Trichoderma* have been known as biocontrol agents of plant diseases since the 1930s (Weindling, 1932). In the 1990s they became increasingly used in agriculture as biofungicides. Potential mechanisms of fungal antagonism include mycoparasitism, caused by the action of cell-wall degrading enzymes, antibiosis, caused by production of antibiotics, competition for space and nutrients through rhizosphere competence, facilitation of seed germination and plant growth via releasing important minerals and trace elements from soil and induction of defense responses in plants (Harman, 2000). These fungi form a physical bond with the root system of plants, establishing themselves in the rhizosphere (root zone) and thereby preventing other pathogens from colonizing the soil. The fungi are known to increase the rate of plant growth and development, and are able to cause the production of more robust roots. This bond and steady growth of *T. harzianum* within the root system form a physical barrier to plant pathogens. The fungus also feeds on excess nutrients left unused by the root system, which would otherwise provide a food source for incoming pathogens. *T. harzianum* does not interfere with the mycorrhizal activity of *Rhizobium* spp. (a common nitrogen-fixer). It releases chitinases, specific enzymes that denature chitin, to break down the cell wall of fungal pathogens in the soil (Waghunde et al., 2016).

T. harzianum has been formulated as the biofungicide RootShield® WP (*T. harzianum* T - 22 KRL-AG2 10^7 CFU/g 1.15%) and is used against many soil-borne pathogens, including *Pythium* spp., *Rhizoctonia* spp., *Fusarium* spp. and other fungi, infecting many wooden and herbaceous plants (corn, soybeans, potatoes, tomatoes, cotton etc.). This product is registered for application on seeds or roots prior to planting. It is mostly used in nurseries and becomes active when temperatures are above 10°C (Anonymous, 2014). This biofungicide is not adequate for application to sugarcane, rice, mushroom, kiwi, tobacco, barley, oats, lemon, apple and chickpea. For drench application to seeds, 5-15 g of the product should be applied per 3.8 L of water, and so 7.6 m^2 of planting furrows would be watered before covering the seeds with soil. For transplants, 5-15 ml of formulation should be mixed with 3.8 L of water and applied at 110-220 ml per plant at transplanting. For new plant beds, 5-15 g of the product should be mixed with 3.8 L of water and applied to 2.3 m^2. For established beds or potted plants, the amount of 5-15 g of biofungicide should be mixed with 3.8 L of water per 2.3 m^2 of plant bed or 110-220 ml per 10-20 cm pot. RootShield® is designed for application to seeds, plant roots and soil for control of plant root diseases and not for spray application to foliage or fruit (Akladious and Abbas, 2014). Sardar Eco Green WP contains *T. harzianum* 2%, 2 x 10^6 CFU/g. It should be applied as 10% paste or slurry to seeds at 5-10 g/kg of seeds, or to seedlings as coating and/or root dipping. Soil application rate is 2.5 kg/ha in 25-50 kg farm yard manure for soil-borne diseases. Additionaly, it can be used as a foliar spray at the rate of 2.5 kg/ha for air-borne diseases (Kumar et al., 2017). The biofungicide Bio-Cure-F contains *T. viride* in liquid (1 x 10^9 per ml) and powder (2 x 10^6 per g) formulation. Rates of application in greenhouse potting mix or soil drench are 1000 ml or 2000 g/100 L. It should be applied at a rate of 200 ml or 200 g per cubic meter of greenhouse potting mix, soil or planting beds, media or substrate. Bulbs should be dipped in a suspension of 10 ml or 20 g/L prior to planting. For seed dressing, 6 ml or 8 g of formulation should be added to 1 kg of seeds and used along with appropriate stickers/wetting agents to coat the seeds. For seedling treatment, an amount of 10 ml or 20 g/L of water should be applied. Roots of the seedlings should be dipped in the solution (Mutawila et al., 2016).

Pythium oligandrum Drechler. The biofungicide Polyversum™ based on *Pythium oligandrum* ATCC 38472, 10^6 oospores/g, was registered in the USA in 2007 and in the EU and China in 2009. In 2011 its registration was expanded to include cereals. In 1993 it was authorized as a seed treatment compound. *P. oligandrum* is a parasite of a number of plant pathogenic fungi: *Botrytis cinerea, Fusarium oxysporum* f. sp. *radices-lycopersici, Pythum ultimum, Rhizoctonia solni, Verticillium albo-atrum* and *Verticillum dahliae.* It has been registered against blackleg (*Leptoshaeria maculans*) and sclertotinia stem rot (*Sclerotinia sclerotiorum*), and against fusariosis in cereals. It is applied at a concentration of 0.05% (Filajdić et al., 2006). Rekanović et al. (2007) reported significant results on the use of *Pythium oligandrum* for controlling *Verticillium* spp. in vegetables. However, it was significantly less effective than the standard fungicide Ronilan-DF against *B. cinerea* on raspberries (Filajdić et al., 2006) or copper oxychloride against *Phomopsis viticola* on grapevine (Latinović et al., 2005). Nevertheless, it is a biological product and almost entirely free of any toxicological or ecotoxicological restrictions. It also offers a significant advantage for widespread usage in plant protection (Rekanović et al., 2007).

Antagonistic bacteria. Most of the bacterial strains commercially used as biopesticides belong to the genera *Bacillus* and *Pseudomonas* (Fravel et al., 2005). Mechanisms that are involved in plant disease suppression by bacteria include siderophore-mediated competition for iron, consumption of pathogen stimulatory compounds, production of antifungal compounds or lytic enzymes, and induction of systemic resistance (Chin-A-Woeng et al., 2003).

***Pseudomonas* spp.** Species in the genus *Pseudomonas* are the most extensively studied group of biocontrol bacteria because they have many traits that make them effective biocontrol agents of plant pathogens (Chin-A-Woeng et al., 2003). They are abundant in nature, able to grow on a wide variety of substrates, have a high growth rate, grow at relatively low temperatures and have a variety of mechanisms to suppress growth of other microorganisms (Weller, 2007). Bacterial isolates stimulating edible mushroom formation have been identified as *P. putida* (Hayes et al., 1969; Rainey et al., 1990).

Pseudomonas fluorescens A506 strain has been formulated as the biofungicide BlightBan A506 71% against *Erwinia amylovora* fire blight on apple and pear, and for suppression of frost damage on cherry, apple, pear, almond, peach, tomato, potato and strawberry in commercial agriculture, using foliar application (Anonymous, 2014) with the following dilution rates: as a concentrate – 190-380 L/0.4 ha, or as a dilute spray 760-1300 L/0.4 ha. The mode of action of this strain is competition with fire blight bacteria for nutrients. Such competition keeps the number of harmful bacteria low enough to avoid severe infection. Additionally, BlightBan is used for reduction of fruit russet, caused by russet-inducing bacteria. For best results in frost protection, initial spray should be applied when plants are at specific stages of development: tomato and potato at the two true-leaf expansion stage; strawberry at first bloom; almond, apricot, blueberry, peach, nectarine when first bloom is at about 10% fully-opened; apple, pear and cherry when green tips are 1 cm long. Treatments should be repeated as requires, making a total of two to three applications. Uniform application to blossom and leaf tissue is essential. For fire blight and fruit russet reduction, this product should be used as part of an integrated fire blight suppression program (Stockwell and Stack, 2007). This product improves suppression with any existing antibiotic spray program. When it is used as an aid to suppress fire blight and fruit russet in pears and apples, the application amount should be 150 g/100 L of water of the product as a concentrate or 200 g/200 L as a dilute spray. The first application should be made at early bloom, followed by a second one at full bloom, and a third application at post petal fall for pear and apples with 75 g/100 L of concentrate or 100 g/200 L of dilute spray. The product should be applied to alternate rows with the first application at early bloom followed by the second application to the untreated rows one week later. The third and fourth applications should use the same rates, applied to alternating rows at first petal fall-full bloom. For the final two applications, the same rates and method should be used as for the final applications at retail bloom for pear or post petal fall for apple. The product can be integrated with all conventional products for disease control other than copper (Stockwell et al., 2010).

***Bacillus* spp.** produce spores that are resistant to various physical and chemical treatments, such as heat, desiccation, UV radiation and organic solvents. They are known to produce an array of secondary metabolites, such as antibiotics, cell-wall degrading enzymes and antifungal volatile substances. Specified activities indicate that *Bacillus* spp. strains could be efficient biological control agents against a wide range of plant and edible mushroom pathogens, both fungal and bacterial (Kim & Chung, 2004; Leelasuphakul et al., 2006; Milijačević et al., 2017; Stanojević et al., 2016). The most probable mode of action of *Bacillus* species is antibiosis as they are known to produce a vast amount of antimicrobial secondary metabolites (Edwards et al., 1994). One of the most studied microorganisms, *B. subtilis*, can utilize 4-5% of the genome for the production of antibiotics (Stein, 2005), while in *B. amyloliquefaciens*, even a greater part of the genome participates in biosynthesis of these molecules (Chen et al., 2009). Additionally, *B. pumilus* and species belonging to *B. subtilis* group are safe for the environment and harmless to human health and are generally recognized as safe (GRAS) organisms (FDA, 1999). The biofungicide Serenade® WP, based on *B. subtilis* (Ehrenberg) Cohn, strain QST 713, has been registered for use against many plant pathogens: *Alternaria* spp., *Botrytis* spp., *Colletotrichum* spp., *Erysiphe* spp., *Monilinia* spp., *Mycosphaerella* spp., *Phomopsis viticola, Phytophthora infestans, Phytophthora parasitica, Podosphaera* spp., *Sclerotinia* spp., *Sphaerotheca* spp., *Uncinula necator* etc. in some western European countries, e.g., Great Britain, Ireland, France, Italy, Germany, Switzerland and Turkey, while a registration process is under way in Spain, Portugal and Greece (Lahlali et al., 2013; Serrano et al., 2013). The biofungicide Real IPM *Bacillus subtilis* is registered in Kenya, and F-stop in Serbia for control of powdery mildew in roses. In mushroom industry, fungicides based on *B. subtilis* can be applied for spawn treatment during spawning: 6 kg of the fungicide Serenade® WP per 1 tone of spawn, and 3-6 g per 1 m^2 for casing treatment, initially four days after casing, and then between all following flushes (Kosanović et al., 2013; Milijašević-Marčić et al., 2017).

Serenade® WP showed satisfactory efficiency against *Cladobotryum* spp., *T. harzianum* and *T. aggressivum* f. *euopaeum* from Serbia *in vivo* (Potočnik et al., 2010; Stanojević et al., 2016; Milijašević-Marčić et al., 2017). It was permitted for treatment of mushroom compost against *T. aggressivum* f. *europaeum* in France after extensive testing was conducted (Védie & Rousseau, 2008). The authors applied 15 g per 100 kg of mushroom compost. Jazz® WP with the same *B. subtilis* QST713 strain is now registered in the USA and Canada against *T. harzianum.* It is applied as: mushroom spawn grains treatment, 3 g of the biofungicide thoroughly mixed with 50 g of gypsum, limestone or chalk, followed by coatin spawn grains (two units) and mixing into the mushroom growing substrate (m^2 of bed surface); mushroom growing supplement, 3 g of the biofungicide thoroughly mixed with 50 g of gypsum, limestone or chalk, followed by coating the supplement (approximately 1800 g) and mixing the supplement into the mushroom growing substrate (m^2 of bed surface); mushroom growing beds treatment, 3 g of the biofungicide Jazz in 0.8 L of irrigation water as a drench to 1 m^2 of bed surface at casing before the first flush (the second day after casing), between the first and second flushes and between the second and third flushes, according to disease pressure. If higher disease pressure is anticipated, it is recommended to use higher dosage. Kosanović et al. (2013) inferred that the biofungicide based on *B. subtilis* QST 713, applied at a rate of 1 g per m^2 of casing soil, demonstrated considerable effectiveness in preventing green mold disease symptoms, compared to tea tree oil, although lower then the fungicide prochloraz. Antagonism was observed in simultaneous application of prochloraz with two biofungicides, for control of *T. harzianum*, and *B. subtilis* showed lower antagonism than tea tree oil to fungicide (Kosanović et al., 2013).

However, improvements in crop yield after application of biocontrol agents have been reported. Tautorus and Townsley (1983) found that a *Bacillus* sp. added to compost inoculated with *Chetomium olivaceum* significantly enhanced the mycelial development of button mushroom yield and efficacy in olive green mold control. Nagy et al. (2012) found that *B. subtilis* exhibited a positive impact on the spawn run of oyster mushroom in the substrate. They also found that a *B. amyloliquefaciens* strain, effective *in vivo* against *Trichoderma pleurotum*, the causal agent of green mold on oyster mushroom, improved crop yield in uninoculated bags of oyster mushrooms by 10% (Gong et. al., 2015).

Application rates should be 100 ml per 100 m^2 for vegetables and flowers, 150 ml per 100 m^2 for lawns and sod fields, and 200 ml in 10 L water for all plant parts in orchards. All treatments should be repeated every seven days. The biofungicide Serenade® based on *B. subtilis* QST713, applied at planting as soil treatment (rather than seed treatment), stimulates the growth of tomato and other cultivated plants. Serenade® Opti is registrated as a fungicide and bactericide at a concentration of 26.2%. Another formulation of the same *B. subtilis* strain QST713 in the biofungicide Rhapsody® WP is registered for lawns and sod fields. A broad spectrum of activity of *B. subtilis* has been registered, including *Botrytis* spp., *Alternaria* spp., *Sclerotinia* spp., *Phytophthora* spp., *Peronospora* spp., *Venturia* spp., *Monilia* spp., *Cercospora* spp., *Septoria* spp. and other pathogens. Additionally, bees can facilitate the spread of *B. subtilis* spores among flowers (Chen et al., 2012). Application at the flowering stage is a single application at a rate of 2-3 kg/ha, while 2-3 successive applications are requires for 1 kg/ha rate during flowering. It can be applied under conditions favorable for infection and at appearance of secondary blossoms (in pear), at the application rate of 2-3 kg/ha, also in tank mixture with copper-based products, both at a reduced rate (Serenade at 1.5 kg/ha). It can also be applied as foliar treatment by conventional spray equipment (Anonymous, 2014). *B. subtilis* is recognized by the relevant industries as a resistance management tool, having low resistance risk mode of action designated by FRAC (2009 update). The primary target is cell membrane, not a protein, and it is not likely for a simple mutation to happen that can make cell resistant strains (Stanojević et al., 2016).

Antagonistic Actinomycetes

Actinomycetes are prokaryiotes with a hyphal morphology. They are soil microorganisms which are active in decomposition of plant tissues and recycling of carbon and nitrogen (Taechowisan et al., 2003). They are widely distributed in nature and have the capacity to synthesize many biologically active secondary metabolites, such as antibiotics, anti-parasitic and enzyme inhibitors. Several properties associated with actinomycetes might explain the ability of some of them to act as biocontrol tools. Those properties include an ability to colonize plant surface, antibiosis against plant pathogens, synthesis of particular extracellular proteins, and degradation of phytotoxins (Doumbou et al., 2017).

The members of the genus *Streptomyces* have been found to produce more than a thousand secondary metabolites with mostly insecticidal and herbicidal activities (Tanaka & Omura 1993). Additionaly, actinomycetes produce a variety of antibiotics that are of special interest due to their diverse biological activities, such as antibacterial, antifungal and antiviral features. For example, kasugamycin (from *Streptomyces kasugaensis*) is a bactericidal and fungicidal metabolite which acts as an inhibitor of protein biosynthesis in microorganisms but not in mammals, and its toxicological properties are excellent (Doumbou et al., 2017). Hokko Chemical Industries has developed a production process to market the systemically active kasugamycin for control of rice blast *Pyricularia oryzae* Cavara and bacterial *Pseudomonas* diseases in several crops. Additionally, polyoxins B and D isolated from *Streptomyces cacaoi* var. *asoensis* have a mode of action that makes them very acceptable with regard to environmental considerations. They interfere with the fungal cell wall synthesis by specifically inhibiting chitin synthase (Endo & Misato, 1969). Polyoxin B has found application against a number of fungal pathogens in fruits, vegetables and ornamentals. Polyoxin D is marketed by several companies to control rice sheath blight caused by *Rhizoctonia solani* Kùhn (Doumbou et al., 2017). Moreover, Validamycin A was found to be a prodrug which is converted within the fungal cell to validoxylamine A, an extremely strong inhibitor of trehalase (Kameda et al., 1987). This mode of action gives validamycin A a favorable biological selectivity because vertebrates do not depend on the hydrolysis of disaccharide trehalose for their metabolism. The antifungal metabolite mildiomycin from a culture of *Streptoverticillium rimofaciens* (Iwasa et al., 1978) is strongly active against several powdery mildews on various crops (Harada & Kishi 1978), acting as an inhibitor of the fungal protein biosynthesis. It is noteworthy that two-thirds of the commercially obtained antibiotics have been isolated from actinomycetes. It is conceivable that a formulated supplement of microbial origin could act as a slowly released nutrient to increase mushroom yield during the later stages of cropping (Fermor & Grant, 1985). It was found that actinomycetes added to mushroom substrate enhanced mushroom yield 10-32% (Jarak et al., 2000), and that would be the factor of greater resistance of *A. bisporus* to *T. aggressivum,* the main compost competitor to mushroom mycelia.

***Streptomyces* spp.** A prime example of *Streptomyces* biocontrol agent is *Streptomyces griseoviridis* strain K61. This strain, originally isolated from light coloured *Sphagnum* peat (Tahvonen, 1982a, 1982b) has been reported to be antagonistic to a variety of plant pathogens including *Alternaria brassicola*, *Botrytis cinerea*, *Fusarium avenaceum*, *F. Culmorum*, *Pythium debaryanum*, *Phomopsis sclerotioides*, *Rhizoctonia solani* and *Sclerotinia sclerotiorum*. *Streptomyces griseoviridis* strain K61 is used in root dipping or growth nutrient treatment of cut flowers, potted plants, greenhouse cucumbers, and various other vegetables (Mohammadi & Lahdenpera, 1992). Mycostop™ (developed by Kemira Oy) is a biofungicide that contains *S. griseoviridis* as its active ingredient. This product is available in the United States (Cross & Polonenko, 1996) and Europe (Tahvonen, 1982a).

In addition, *Streptomyces lydicus* WYEC 08 which is active in the biofungicides Actinovate and Actono-iron has been registered against turf diseases, including *Pythium* spp., *Rhizoctonia* spp., *Fusarium* spp., *Sphaerotheca* spp., *Botrytis* spp., *Alternaria* spp. and it is applied as soil drench or foliar treatment (Anonymous, 2014). Its recommended rates are 0.05% and the double high 0.1% (Young and Crawford, 1995; Zeng et al., 2012).

Several scientific publications have indicated that actinomycete species are capable of effectively controlling fungal and bacterial plant pathogens. In most cases, the levels of biocontrol achieved by the various actinomycetes in laboratory or controlled-environment studies are sufficient to suggest that they could provide reliable and effective control, alternative or complementary to chemical pesticides. Therefore, a few plant pathogens have been controlled successfully by actinomycetes species, but many attempts to develop biocontrol formulations have met with problems in practice. In order to develop actinomycete biocontrol agents for commercial use, the consistency of their performance must be improved. Accomplishing this will require research into many diverse areas because biological control is the culmination of complex interactions among the host, pathogen, antagonist, and environment (Doumbou et al., 2017).

Challenges in Microbial Biofungicides Implementation

Microbial fungicides have some limitations compared to the features that chemical products have. They do not inhibit mycelial growth when they are applied a few days before the pathogen. Therefore, biopesticides have to be applied as a preventive measure prior to pathogen infestation, but just before the parasite/pathogen emergence. They act more slowly and have short-time effects in comparison with chemical pesticides. Microbial units must be present in considerable amounts and therefore repeated applications are recommended. Biofungicides are often much more expensive than chemical compounds. Usually, they have to be applied at higher doses than chemical standards and treatments repeated within 7-10 days. Furthermore, some of them may provoke allergic reactions or breathing problems, such as *Streptomyces* spp., *Trichoderma* spp., *Pythium* spp., *Talaromyces* spp., *Coniothyrium* spp., etc. *Ampelomyces* could pose a potential risk for some micorrhizal and saprobic fungi, soil-born beneficial bacteria, insects and other non-target organisms. There are climate limitations for applying these products, such as low temperature and low humidity, etc. Their storage includes some restrictions and they need to be kept in cool and dark chembers. The type of biofungicide formulation and carriers are important for their successful application. In many cases, solid formulations are preferable, compared with concentrate emulsions. Some of them show phytotoxic effects, such as *Trichoderma* spp., *Gliocladium* spp., *Rhizobium* spp., etc. Their activity could be diminished by other competitive microorganisms in the rhizosphere, such as *T. harzianum.* Antagonistic reactions with chemical pesticides have been reported, and many of them should be applied independent of such chemicals (Kosanović et al., 2013).

Moreover, biofungicides maintain diseases below threshold. Above the threshold, profits are reduced due to cost increases and reduced quality and storability. As a consequence, disease control (biological or other) must be extremely effective and able to compete with all available fungicides. In practice, a biological product is unlikely to show all the characteristics of modern fungicides, such as applicability under diverse climatic conditions, high leaf adherence and rapid eradication capability. Unless growers can receive market premium for some agricultural products produced with less pesticides or implemented IPM practices, they will not replace chemical fungicides with biological ones. Despite these hurdles, the concept of "biosanitation" remains attractive to growers for several reasons. By integrating a biofungicide that reduces initial inoculum into production systems, they reduce the risk of disease control failure, which may occur under high disease pressure. Furthermore, they increase the potential for significantly reducing the number of fungicide applications. By spraying crops in early season, the growers can still rely on their conventional program for disease management. Growers retain the same fungicide program in early season, following an application of a biofungicide but as they become more accustomed to the impacts of the product, they may decide to eliminate unnecessary chemical treatments. They may not need to use it every year, but in seasons when disease pressure becomes very high, a biofungicide will enable them to bring the inoculum to a manageable level. At the end of the season, fungicide residues in crops are generally not an issue. A single application allows integration with other practices and selection of the most ideal weather conditions. Biosanitation is easily combined with other sanitation practices to reduce inoculum. It will not be cost-effective unless it reduces disease management costs or yield losses during the growing season. To attain a significant reduction in the cost of disease control, biosanitation practices and sanitation are generally combined with a delayed fungicide program (Vincent et al., 2007).

To conclude, inventing and developing biological methods for plant disease control is highly important, considering pathogen resistance development to chemicals, their harmful impact on the environment and human health, as well as increase in production costs. Pollution control legislation and environmental awareness are becoming more and more rigorous, and new technologies are implemented in agriculture that ensure production with the least possible harmful effects (Todorović et al., 2016). However, in order to conceive more effective biological control strategies in the future, it is critical to carry out expand research to some less studied aspects of biocontrol, including the development of novel formulations, understanding the impact of environmental factors on biocontrol agents, mass production of biocontrol microorganisms and the use of biotechnology and nano technology for improving biocontrol mechanisms and strategies. Future outlooks of biocontrol of plant diseases seem to be promising and with a growing demand for biocontrol products among growers, it is possible to use biological control as an effective strategy to manage plant diseases, increase yield, protect the environment and biological resources, and approach a sustainable agricultural system (Hasan et al., 2013).

Acknowledgments

This study was funded by the Ministry of Education, Science and Technological Development of the Republic of Serbia, project TR 31043.

References

Akladious, S.A. and Abbas, S.M. (2014). Application of *Trichoderma harzianum* T22 as a biofertilizer potential in maize growth. *J. Plant Nutrit.* 37 (1), 30-49.

Anonymous (2014). UC IPM. *Green Bull.* 4 (2), 1-6.

Aravind, R.; Eapen, S.J.; Kumar, A.; Dinu, A. and Ramana, K.V. (2010). Screening of endophytic bacteria and evaluation of selected isolates for suppression of burrowing nematode (*Radopholus similis* Thorne) using three varieties of black pepper (*Piper nigrum* L.). *Crop Protect. 29*, 318–324.

Aravind, R.; Kumar, A.; Eapen, S.J. and Ramana, K.V. (2009). Endophytic bacterial flora in root and stem tissues of black pepper (*Piper nigrum* L.) genotype: isolation, identification and evaluation against *Phytophthora capsici*. *Lett. Appl. Microbiol.* 48, 58–64.

Bakker P.A.H.M.; Ran, L.X.; Pieterse, C.M.J. and van Loon, L.C. (2003). Understanding the involvement of rhizobacteria-mediated induction of systemic resistance in biocontrol of plant diseases. *Can. J. Plant Pathol.* 25, 5–9.

Basumatary, M.; Kumar, A.; Dutta, B.; Singha, D.M. and Das, N. (2015). Some *in vitro* observations on the biological control of *Sclerotium rolfsii*, a serious pathogen of various agricultural crop plants. *IOSR J. Agric. Veterin. Sci.* 8 (2), 87-94.

Benitez, M.S. and McSpadden Gardener, B.B. (2009). Linking sequence to function in soil bacteria sequence-directed isolation of novel bacteria contributing to soilborne plant disease suppression. *Appl. Environ. Microbiol.* 75, 915–924.

Berg, G. (2007). Biologcial control of fungal soilborne pathogens in strawberries. In: Chincholkar, S.B., Mukerji, K.G. (eds) *Biological control of plant diseases*. The Haworth Press, Inc, Birghamton.

Berg, G.; Marten, P.; Minkwitz, A. and Brückner, S. (2001). Efficient biological control of fungal plant diseases by *Streptomyces* sp. DSMZ 12424. *J. Plant Dis. Protect.* 108, 1–10.

Berendsen, R.L.; Kalkhove, S.I.C.; Lugones, L.G.; Baars, J.J.P.; Wösten, H.A.B. and Bakker, P.A.H.M. (2012). Effects of fluorescent *Pseudomonas* spp. isolated from mushroom cultures on *Lecanicillium fungicola*. *Biol. Control* 63, 210-221.

Berić, T.; Kojić, M.; Stanković, S.; Topisirović, Lj.; Degrassi, G.; Myers, M.; Venturi, V. and Fira Dj. (2012). Antimicrobial Activity of *Bacillus* sp. Natural Isolates and Their Potential Use in the Biocontrol of Phytopathogenic Bacteria. *Food Technol. Biotechnol.* 50 (1), 25–31.

Bolwerk, A.; Lagopodi, A.L.; Lugtenberg, B.J.J. and Bloemberg, G.V. (2005). Visualization of interactions between a pathogenic and beneficial *Fusarium* strain during biocontrol of tomato foot and root rot. *Mol. Plant–Microbe Interact.* 18, 710–721.

de Boer, W.; Folman, L.B.; Summerbell, R.C. and Boddy, L. (2005). Living in a fungal world: impact of fungi on soil bacteria niche developments. *FEMS Microbiol. Rev.* 29, 795–811.

Chaiharn, M.; Chunhaleuchanon, S. and Lumyong, S. (2009). Screening siderophore producing bacteria as potential biological control agent for fungal rice pathogens in Thailand. *World J. Microbiol. Biotech.* 25, 1919–1928.

Chen, X.H.; Koumoutsi, A.; Scholz, R.; Schneider, K.; Vater, J.; Süssmuth, R.; Piel, J. and Borriss, R. (2009). Genome analysis of *Bacillus amyloliquefaciens* FZB42 reveals its potential for biocontrol of plant pathogens. *J. Biotech.* 140 (1), 27-37.

Chen, Y.; Yan, F.; Chai, Y.; Liu, H.; Kolter, R.; Losick, R. and Gun, J-H. (2012). Biocontrol of tomato wilt disease by *Bacillus subtilis* isolates from natural environments depends on conserved genes mediating biofilm formation. *Environ. Microbiol.* 15 (3), 848-864.

Chin-A-Woeng, T.F.C.; Bloemberg, G.V. and Lugtenberg, B.J.J. (2003). Phenazines and their role in biocontrol by *Pseudomonas bacteria. New Phytologist* 3 (157), 503-523.

Cross, J.V. and Polonenko, D.R. (1996). An industry perspective on registration and commercialization of biocontrol agents in Canada. *Can. J. Plant Pathol.* 18, 446-454.

Dimkić, I.; Živković, S.; Berić, T.; Ivanović, Ž.; Gavrilović, V.; Stanković, S. and Fira, Dj. (2013). Characterization and evaluation of two *Bacillus* strains, SS-12.6 and SS-13.1, as potential agents for the control of phytopathogenic bacteria and fungi. *Biol. Control* 65,312-321.

Duijff, B.K.; Pouhair, D.; Olivain, C.; Alabouvette, C. and Lemanceau, P. (1998). Implication of systemic induced resistance in the suppression of *Fusarium* wilt of tomato by *Pseudomonas fluorescens* WCS417r and nonpathogenic *Fusarium oxysporum* Fo47. *Eur. J. Plant Pathol.* 104, 903–910.

Doumbou, C.L.; Akimov, V. and Beaulieu, C. (1998). Selection and characterization of microorganisms utilizing thaxtomin A, a phytotoxin produced by *Streptomyces scabies*. *Appl. Environ. Microbiol.* 44, 4313-4316.

Edwards, S.G.; McKay, T. and Seddon B. (1994). Interaction of *Bacillus* species with phytopathogenic fungi-Methods of analysis and manipulation for biocontrol purposes. In: Blakeman, J.P. and Williamson, B, (Eds.), Ecology of Plant Pathogens (pp. 101-118). Wallingford, Oxon, UK: CAB International.

Endo, A. And Misato, T. (1969). Polyoxin D, a competitive inhibitor of UDP-N-acetylglucosaminyltransferase in *Neurospora crassa. Biochem. Biophys. Res. Commun.* 37, 718-722.

Fermor, T.R. and Grant, W.D. (1985). Degradation of fungal and actinomycete mycelia by *Agaricus bisporus*. *J. Gen. Microb.* 131, 1729-1734.

Filajdić, N.; Rekanović, E.; Tanović, B.; Stević, M. and Vukša, P. (2006). Efikasnost biofungicida Polyversum™ u suzbijanju *Botrytis cinerea* Pers. na plodovima maline [Efficacy of the Biofungicide Polyversum™ in Controlling *Botrytis cinerea* Pers. On Raspberry]. *Pesticidi i fitomedicina* [*Pesticides and Phytomedicine*] 21, 311-316.

Fokkema, N.J. (1973). The role of saprophytic fungi in antagonism against Drechslera sorokiniana (*Helminthosporium sativum*) on agar plates and on rye leaves with pollen. *Physiol. Plant. Pathol*, 3, 195- 205.

Fravel, D.R. (2005). Commercialization and implementation of biocontrol. *Ann. Rev. Phytopathol.* 43, 337-359.

Fuchs, J.-G.; Moenne-Loccoz, Y. and Defago, G. (1997). Nonpathogenic *Fusarium oxysporum* strain Fo47 induces resistance to *Fusarium* wilt in tomato. *Plant Dis.* 81, 492– 496.

Galindo, E.; Serrano Carreón, L.; Gutièrrez, C.R.; Allende, R.; Balderas, K.; Patino, M.; Trejo, M.; Wong, M.A.; Rayo, E.; Isauro, D. and Jurado, C. (2013). The challenges of introducing a new biofungicide to the market: A case study. *Electronic J Biotechnol.* doi 10.2225/vol16-issue3-fulltext-6.

Giacomodonato, M.N.; Pettinari, M.J.; Souto, G.I.; Mendez, B.S. and Lopez, N.I. (2001). A PCR based method for the screening of bacterial strains with antifungal activity in suppressive soybean rhizosphere. *World J. Microbiol. Biotech.* 17, 51–55.

Gong, A-D.; Li, Q-S.; Song, X-S.; Yao, W.; He, W-J.; Zhang, J-B. and Liao, Y-C. (2015). Antagonistic Mechanism of Iturin A and Plipastatin A from *Bacillus amyloliquefaciens* S76-3 from Wheat Spikes against *Fusarium graminearum. PloS One* doi: 10(2):e0116871.

Haas, D. and Defago, G. (2005). Biological control of soil-borne pathogens by fluorescent pseudomonads. *Nat. Rev. Microbiol.* 3, 307–319.

Haas, D. and Keel, C. (2003). Regulation of antibiotic production in root-colonizing *Pseudomonas* spp. and relevance for biological control of plant disease. *Annu. Rev. Phytopathol.* 41, 117–153.

Harada, S. and Kishi, T. (1978). Isolation and charactrization of mildiomycin, a new nucleoside antibiotic. *J. Antibiot.* 31, 519.

Harman, G.E.; Howell, C.H.; Vitero, A.; Chet, I. and Lorito, M. (2004). *Trichoderma* species opportunistic, avirulent plant symbionts. *Nature Rev. Microbiol.* 2, 43–56.

Harman, G.E. (2000). Myths and dogmas of biocontrol. Changes in perceptions derived from research on *Trichoderma harzianum* T-22. *Plant Dis.* 84, 377-393.

Hasan, S.; Gupta, G.; Anand, S. and Chaturvedi, A. (2013). Biopotential of microbial antagonists against soilborne fungal plant pathogens. *Internat. J. Agric. Food Sci. Technol. (IJAFST)* 4 (2), 37-39.

Hayes, W.A.; Randle, P.E. and Last, F.T. (1969). The nature of the microbial stimulus affecten sporophore formation in *Agaricus bisporus* (Lange) Sing. *Ann. Appl. Biol.* 64, 177-187.

Hoster, F.; Schmitz, J.E. and Daniel, R. (2005). Enrichment of chitinolytic microorganisms: isolation and characterization of a chitinase exhibiting antifungal activity against phytopathogenic fungi from a novel *Streptomyces* strain. *Appl. Microbiol. Biotechnol. 66*, 434–442.

Iwasa, T.; Suetomi, K. And Kusuka, T. (1978). Taxonomic study and fermentation of producing organism and antimicrobial activity of mildiomycin. *J. Antibiot.* 31, 511-518.

Jarak, M.; Govedarica, M.; Milošević, N. and Klokočar-Šmit, Z. (2000). Primena aktinomiceta u proizvodnji šampinjona. *Zbornik radova Naučnog Instituta za ratarstvo i povrtarstvo* [Application of actinomycetes in the production of button mushrooms. *Proceedings of the Institute of Field and Vegetable Crops*], Novi Sad 34, 77-81.

Kameda, Y.; Asano, N.; Yamaguchi, T. and Matsui, K. (1987). Validoxylamines as trehalase inhibitors. *J. Antibiot.* 40, 563-565.

Kamilova F.; Validov S.; Azarova T.; Mulders I.; Lugtenberg B.J.J. (2005). Enrichment for enhanced competitive plant root tip colonizers selects for a new class of biocontrol bacteria. *Environ. Microbiol.* 7, 1809–1817.

Kim, P. and Chung KC. (2004). Production of an antifungal protein for control of *Colletotrichum lagenarium* by *Bacillus amyloliquefaciens* MET0908. *FEMS Microbiol. Letters.* 234, 177-183.

Kloepper, J. W.; Ryu, C.-M. and Zhang, S.A. (2004). Induced systemic resistance and promotion of plant growth by *Bacillus* spp. *Phytopathology* 94,1259-1266.

Kosanović, D.; Potočnik, I.; Duduk, B.; Vukojević, J.; Stajić, M.; Rekanović, E. and Milijašević-Marčić, S. (2013): *Trichoderma* species on *Agaricus bisporus* farms in Serbia and their biocontrol. *Ann. Appl. Biol.* 163, 218-230.

Koumoutsi, A.; Chen, X.H.; Henne, A.; Liesegang, H.; Gabriele, H.; Franke, P.; Vater, J. and Borris, R. (2004). Structural and functional characterization of gene clusters directing nonribosomal synthesis of bioactive lipopeptides in Bacillus amyloliquefaciens strain FZB42. *J. Bacteriol.* 186, 1084–1096.

Kuiper, I.; Bloemberg, G.V. and Lugtenberg, B.J.J. (2001). Selection of a plant–bacterium pair as a novel tool for rhizostimulation of polycyclic aromatic hydrocarbondegrading bacteria. *Mol. Plant–Microbe Interact.* 14, 1197–1205.

Kumar, G.; Maharshi, A.; Patel, J.; Mukherjee, A.; Singh, H.B. and Sarma, B.K. (2017). *Trichoderma:* A Potential Fungal Antagonist to Control Plant Diseases. *SATSA Mukhapatra – Ann.l Technical Issue,* 21, 206-218.

Larkin, R.P. and Fravel, D.R. (1998). Efficacy of various fungal and bacterial biocontrol organisms for control of Fusarium wilt of tomato. *Plant Dis*. 82, 1022–1028.

Lahlali, R.; Peng, G.; Gossen, B.D.; McGregor, L.; Yu, F.Q.; Hynes, R.K.; Hwang, S.F.; McDonald, M. R. and Boyetchko, S.M. (2013). Evidence that the Biofungicide Serenade (*Bacillus subtilis*) Suppresses Clubroot on Canola via Antibiosis and Induced Host Resistance. *Phytopathology* 103, 245-254.

Latinović, N.; Vučinić, Z. and Vukša, P. (2005). Efikasnost biofungicida Polyversum (*Pythim oligandrum* Drechsler) u suzbijanju crne pjegavosti vinove loze [Efficacy of the biofungicide Polyversum (*Pythium oligandrum*) in the control of *Phomopsis viticola*]. *Pesticidi i fitomedicina* [*Pesticides and Phytomedicine*] 20 (1), 37-41.

Leelasuphakul, W.; Sivanunsakul, P.; Phongpaichit. S. (2006). Purification, characterization and synergistic activity of β-1,3-glucanase and antibiotic extract from an antagonistic *Bacillus subtilis* NSRS 89-24 against rice blast and sheath blight pathogens. *Enzyme Microb. Tech.* 38, 990-997.

Leveau, J.H.J.; Gerards, S.; Fritsche, K.; Zondag, G. and van Veen, J.A. (2006). Genomic flank - sequencing of plasposon insertion sites for rapid identification of functional genes. *J. Microbiol. Methods* 66, 276–285.

Loper, J.E.; Kobayashi, D.Y. and Paulsen, I.T. (2007). The genomic sequence of Pseudomonas fluorescens Pf-5: insight into biological control. *Phytopathology* 97, 233–238.

Lugtenberg, B.J.J. and Kamilova, F. (2009). Plant-Growth promoting rhizobacteria. *Ann. Rev. Microbiol.* 63, 541–556.

Milijašević-Marčić, S.; Stanojević, O.; Potočnik, I.: Todorović, B.; Stepanović, M. and Rekanović, E. (2016). Antagonistic activity of *Bacillus* spp. strains against *Pseudomonas tolaasii* and *Pseudomonas agarici.* Book of abstracts III International *Symposium on Biological Control of Plant Bacterial Diseases*, *Belgrade, Serbia*, pp 41.

Milijašević-Marčić, S.; Stepanović, M.; Todorović, B.; Duduk, B.; Stepanović, J.; Rekanović, E. and Potočnik, I. (2017). Biological control of green mould on *Agaricus bisporus* by a native *Bacillus subtilis* strain from mushroom compost *Eur. J. Plant Pathol.* 148(3), 509-519.

Milijašević-Marčić, S. and Todorović, B. (2017). Biological Control of Bacterial Pathogens in Horticultural Systems. In: Biocontrol Agents: Types, Applications and Reserarch Insights (Victor Green, ed.), Nova Science Publishers, Inc., New York, USA, pp. 1-40. ISBN 978-1-53610-553-7.

Milijašević-Marčić, S.; Todorović, B.; Potočnik, I.; Stepanović, M. and Rekanović, E. (2012): First report of *Pseudomonas tolaasii* on *Agaricus bisporus* in Serbia. *Phytoparasitica* 40 (3), 299-303.

Mutawila, C.; Halleen, F. and Mostert, L. (2016). Optimisation of time of application of *Trichoderma* biocontrol agents for protection of grapevine pruning wounds. *Australian J. Grape Wine Res.* 22 (2), 279-287.

Nagy, A.; Manczinger, L.; Tombácz, D.; Hatvani, L.; Gyõrfi, J.; Antal, Z.; Sajben, E.; Vágvõllgyi, C. and Kredics, L. (2012). Biological control of oyster mushroom green mould disease by antagonistic *Bacillus* species. *Biol. Control Fungal Bacterial Plant Pathogens IOBC-WPRS Bull.* 78, 289-293.

Nakaew, N.; Rangjaroen, C. and Sungthong, R. (2015). Utilization of rhizospheric *Streptomyces* for biological control of *Rigidoporus* sp. causing white root disease in rubber tree. *Eur. J. Plant Pathol.* 142, 93–105.

Raaijmakers, J.M.; de Bruijn, I. and de Kock, M.J.D. (2006). Cyclic lipopeptide production by plant-associated *Pseudomonas* spp.: diversity, activity, biosynthesis, and regulation. *Mol. Plant-Microb. Interact.* 19, 699–710.

Raaijmakers, J.M., Vlami, M. and de Souza, J.T. (2002). Antibiotic production by bacterial biocontrol agents. *Antonie van Leeuwenhoek* 81, 537–547.

Pal, K. K. and McSpadden Gardener, B. (2006). Biological control of plant pathogens. *Plant Health Instructor* 2, 1117-1142.

Pliego, C.; Ramos, C.; de Vicente, A. and Cazorla, F.M. (2011). Screening for candidate bacterial biocontrol agents against soilborne fungal plant pathogens. *Plant Soil* 340, 505–520.

Potočnik, I.; Vukojević, J.; Stajić, M.; Rekanović, E.; Stepanović, M.; Milijašević, S. and Todorović, B. (2010). Toxicity of biofungicide Timorex 66 EC to *Cladobotryum dendroides* and *Agaricus bisporus*. *Crop Prot.* 29, 290-294.

Rainey, P.B.; Cole, A.I.J.; Fermor, T.R. and Wood, D.A. (1990). A model system for examining involvement of bacteria in basidiome initiation of *Agacus bisporus*. *Mycol. Res.* 94, 191-4195.

Rekanović E.; Milijašević S.; Todorović B. and Potočnik I. (2007): Possibilities of Biological and Chemical Control of Verticillium Wilt in Pepper. *Phytoparasitica* 35 (5), 436-441.

Romero, D.; Pérez-García, A.; Rivera., M.E.; Cazorla, F.M. and de Vicente, A. (2004). Isolation and evaluation of antagonistic bacteria towards the cucurbit powdery mildew fungus *Podosphaera fusca*. *Appl. Microbiol. Biotechnol.* 64, 263–269.

Shafi, J., Tian, H., Ji, M. (2017). *Bacillus* species as versatile weapons for plant pathogens: a review. *Biotech. Biotechnol. Equipment*, 31(3), 446-459.

Shoresh, M.; Harman, G.E. and Mastouri, F. (2010). Induced systemic resistance and plant responses to fungal biocontrol agents. *Annu. Rev. Phytopathol.* 48, 21-43.

Serrano, L.; Manker, D.; Brandi, F.; Cali, T. (2013). The use of B*acillus subtilis* QST 713 and *Bacillus pumilus* QST 2808 as protectant fungicides in conventional application programs for black leaf streak control. ISHS Acta Horticulturae 986: *VII International Symposium on Banana: ISHS-ProMusa Symposium on Bananas and Plantains: Towards Sustainable Global Production and Improved Use*.

Solanki, M. K.; Singh, R. K.; Srivastava, S.; Kumar, S.; Kashyap, P. L. and Srivastava, A.K. (2013). Characterization of antagonistic-potential of two *Bacillus* strains and their biocontrol activity against *Rhizoctonia solani* in tomato. *J. Basic Microbiol.* 53, 1–9.

Stanojević O.; Milijašević-Marčić S.; Potočnik I.; Stepanović M.; Dimkić I.; Stanković S. and Berić T. (2016). Isolation and identification of *Bacillus* spp. from compost material, compost and mushroom casing soil active against *Trichoderma* spp. *Arch. Biol. Sci. Belgrade* 68 (4), 845-852.

Stein T. (2005). *Bacillus subtilis* antibiotics: Structures, syntheses and specific functions. *Molec. Microbiol.* 56, 845-857.

Stockwell, V.O.; Hohnson, K.B.; Sugar, D. and Loper, J.E. (2010). Control of fire blight by *Pseudomonas fluorescens* A506 and *Pantoea vagans* C9-1 applied as single strains and mixed inocula. *Phytopathology* 100 (12), 1330-1339.

Stockwell, V.O. and Stack, J.P. (2007). Using *Pseudomonas* spp.. for integrated biological control. *Phytopathology* 97, 244-249.

Taechowisan, T.; Peberdy, J.F. and Lumyoung, S. (2003). Isolation of endophytic actinomycetes from selected platns and their antifungal activity. *World J. Microbiol. Biotechnol.* 19, 381-385.

Tahvonen, R. (1982a). The suppressiveness of Finnish light coloured *Sphagnum* peat. *J. Agric. Sci. Finl.* 54, 345-356.

Tahvonen, R. (1982b). Preleminary experiments into the use of *Streptomyces* spp. isolated from peat in the biological control of soil and seedborne disease in peat culture. *J. Agric. Sci. Finl.* 54, 357-369.

Tautorus, T.E. and Townsley, P.M. (1983). Biological Control of Olive Green Mold in *Agaricus bisporus* Cultivation. *Appl. Environ. Microb.* 45 (2), 511-515.

Tanaka, Y. and Omura, S. (1993). Agroactive compounds of microbial origin. *Annu. Rev. Microb.* 47, 57-87.

Todorović, B.; Milijasević-Marčić, S.; Potočnik, I.; Stepanović, M.; Rekanović, E.; Nikolić-Bujanović, Lj. and Čekerevac, M. (2012): *In vitro* activity of antimicrobial agents against *Pseudomonas tolaasii,* pathogen of cultivated button mushroom. *J. Environ. Sci. Heal. B* 47 (3), 175-179.

Todorović, B.; Potočnik, I.; Rekanović, E;, Stepanović, M.; Kostić, M.; Ristić, M. and Milijašević-Marčić, S. (2016). Toxicity of twenty-two plant essential oils against pathogenic bacteria of vegetables and mushrooms *J. Environ. Sci. Heal. B* 47 (3), 175-179.

van Loon, L.C. (2007). Plant responses to plant growth-promoting bacteria. *Eur. J. Plant Pathol.* 119:243–254.

Védie, R. and Rousseau, T. (2008). Serenade biofungicide: une innovation mjeure dans les champignonnières françaises pour lutter contre *Trichoderma aggressivum*, agent de la moisissure verte du compost [Biofungicide Serenade: a major innovation in disease control against *Trichoderma aggressivum*, agent of compost green mold of button mushroom in France]. *La Lettre du CTC* [*Letter of the CTC*] 21, 1-2.

Vincent, C.; Goettel, M.S. and Lazarovits, G. (2007). Biological Control a Global Perspective – Case Studies from Around the World. CABI: Wallingford, UK, pp. 235-236.

Waghunde, R.R.; Shelake, R.M. and Sabalpara, A.N. (2016). Trichoderma: A significant fungus for agriculture and environment. *African J. Agric. Res.* 1952-1965.

Walker, R.; Powell, A.A. and Seddon, B. (1998). *Bacillus* isolates from the spermosphere of peas and dwarf French beans with antifungal activity against *Botrytis cinerea* and *Pythium* species. *J. Appl. Microbiol.* 84, 791-800.

Weindling, R. (1932). *Trichoderma lignorum* as a parasite of other soil fungi. *Phytopathology* 22, 837-845.

Weller, D.M. (2007). Pseudomonas biocontrol agents of soilborne pathogens: looking back over 30 years. *Phytopathology* 97, 250-256.

Yoshida, S.; Hiradate, S.; Tsukamato, T.; Hatakeda, K. and A. Shirata, (2001). Antimicrobial activity of culture filtrate *Bacillus amyloliquefaciens* RC-2 isolated from mulberry leaves. *Phytopathology* 91, 181-187.

Young, W.M. and Crawford, D.I. (1995). Charactrisation of *Streptomyces lydicus* WYEC108 as a potentional biocontrol agent against fungi root and seed rots. *Appl. Envir. Microbiol.* 319-3128.

Zeng, W.; Kirk, W. and Hao, J. (2012). Field management of *Sclerotinia* stem rot of soybean using biological control agents. *Biol. Control* 60, 141-147.

INDEX

A

B

C

D

E

F

G

H

I

L

M

N

O

P

R

S

T

U

V

W

Y